D0816458

Alien Encounters

MYSTERIES OF THE UNKNOWN

Alien Encounters

By the Editors of Time-Life Books

TIME-LIFE BOOKS, ALEXANDRIA, VIRGINIA

CONTENTS

Journeys to Other Worlds

The latter half of the twentieth century has often been called the Space Age, but it might arguably be dubbed the Alien Age as well. Ever since an American businessman reported seeing what he called "saucers" flying over Washington state in 1947, hundreds of thousands of people around the world have announced their own sightings of unidentified flying objects, or UFOs. Of these witnesses, more than three hundred report that they were taken aboard UFOs—many to be subjected to physical examinations by extraterrestrial beings. Still more extraordinary, about thirty of those who claim to have been abducted say that after their physicals they were transported beyond Earth's familiar surroundings to visit other worlds.

Most of the alleged voyagers recall being taken to barren planets with sparse, dimly lit landscapes like the one above. But a few report more exotic destinations. In the following pages, illustrations based on their accounts depict four such unearthly scenes, ranging from a space dock to a museum of human beings.

Among UFO researchers, accounts of otherworldly journeys remain highly controversial. Some students of the UFO phenomenon dismiss all alien abductions as fantasies that taint inquiry into legitimate sightings; others believe that aliens may indeed have taken selected humans on personal cosmic tours. Nor are those the only possibilities. UFO theorists have also suggested the witnesses may have visited an alternate universe—or may have undergone a purely mental voyage, courtesy of alien mind control.

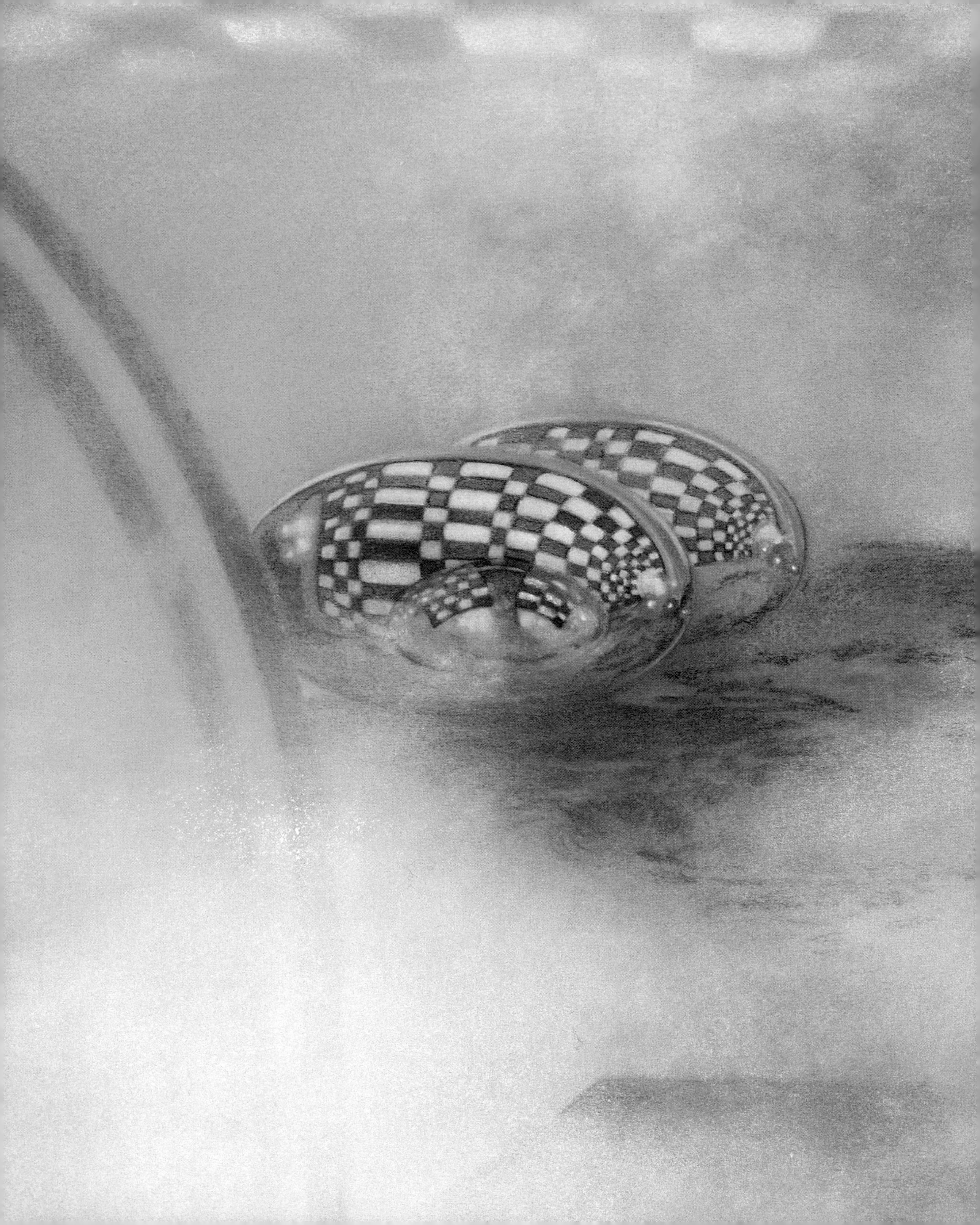

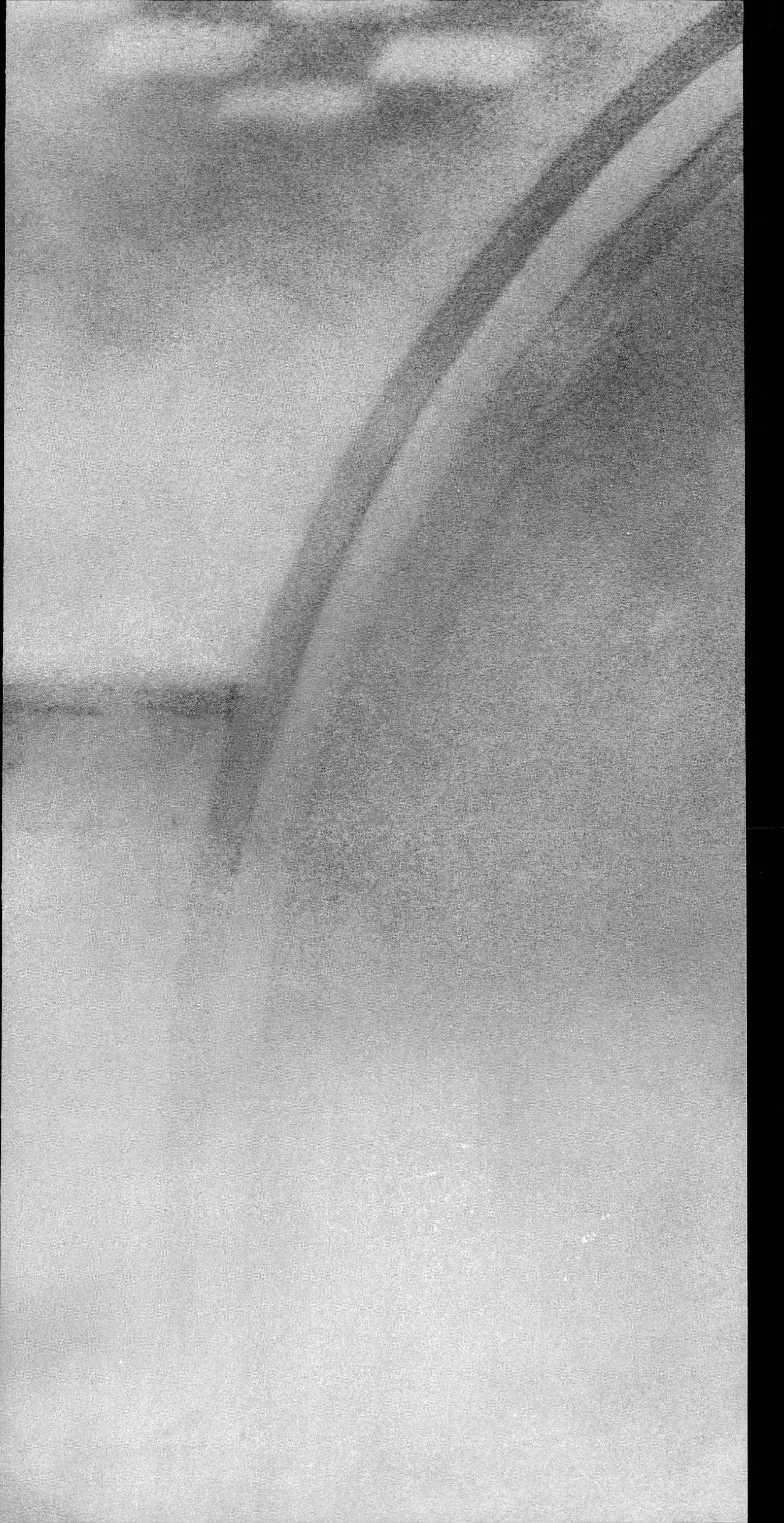

A Glimpse of an Unearthly Hangar

At dusk on the evening of November 5, 1975, a twenty-two-year-old logger named Travis Walton and six of his co-workers were driving home after a long day's work near Heber, Arizona. Suddenly, as the seven men later told the story, they noticed a glowing, twenty-foot disk hovering above a clearing in the forest. According to his companions, Walton jumped out of the truck and ran toward the craft, which shot a blue-green ray of light at him that knocked him unconscious. His friends drove off in terror but returned shortly to help him. By that time, however, both Travis Walton and the UFO had vanished.

When Walton reappeared sometime later, he had an eerie tale to tell. After being knocked out by the UFO beam, he said, he had regained his senses inside the same ship, where he found himself being examined by three short humanoids with large, hairless heads and catlike eyes. Fleeing his captors, he soon encountered an ordinary looking man who silently escorted him out of the UFO and into a huge hangarlike enclosure, depicted at left in an illustration based on Walton's published recollections.

Walton had seemingly been transported off the earth and into an alien space dock filled with other ships and enclosed by a curved checkerboard roof. After passing through the hangar, which he said was filled with fresh air, Walton followed the man into a separate room, where two men and a woman, all human but with a curiously "perfect" appearance, led him to a table. Walton blacked out once again, only to awaken on a road twelve miles from the abduction site. Although he said he could recall only about two hours of his otherworldly voyage, Walton had been missing for five days.

Celestial Ride to Rejection

Forty-year-old oil driller Carl Higdon was apparently expecting nothing more than a good day's hunting in Wyoming's Medicine Bow Forest on the afternoon of October 25, 1974. Spying a bull elk and four females, Higdon took aim at the bull and fired. But, he related afterward, the bullet traveled only fifty feet or so before dropping abruptly to the ground. Higdon went to retrieve it, only to discover that "in the shadow of the trees was this sort of man standing there."

The humanoid figure, who called himself Ausso, must have had a commanding presence, for when he held out a packet of capsules Higdon swallowed one. Ausso then gestured with his right arm, and suddenly Higdon was seated next to his captor inside a transparent cubicle, with all five elk in a cage behind them.

As the cube-shaped vessel took off, Higdon said he observed Earth receding in the distance. Then, almost immediately, they landed on a dark surface, which Higdon surmised was Ausso's planet. As depicted at right, Higdon *(upper corner, near cubicle)* found himself standing near a huge tower that flashed with painfully brilliant light in the gray, foggy air. Not far away, five human beings stood talking, oblivious to Higdon's arrival.

According to Higdon, he had little time to observe the scene, for he was soon whisked off to an examination room within the tower. There Ausso passed a large shield over Higdon's body and told him he was "not any good for what we need." Higdon later speculated that he was rejected because he had had a vasectomy and so could not be used for breeding. Returning to the traveling cube, Higdon soon was back in the forest, two hours after he had fired his gun.

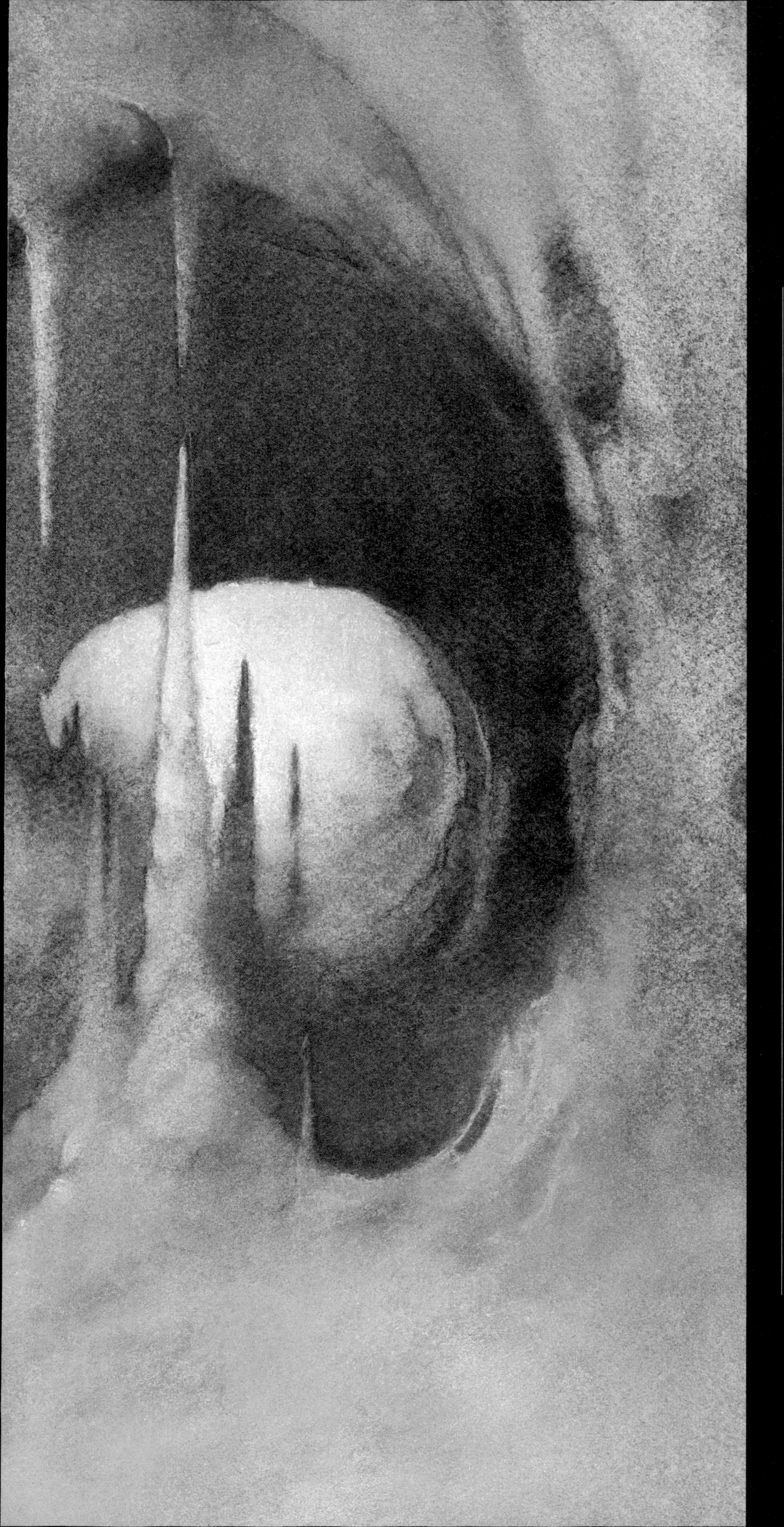

A Museum of Human Specimens

Few otherworldly journeys have been described in such extraordinary detail as the voyages allegedly experienced by a New England woman named Betty Andreasson. Almost completely repressed for years, Andreasson's supposed memories of her cosmic travels were uncovered through hypnosis sessions in 1977 and 1980.

According to those recollections, Andreasson had first been contacted by aliens when she was seven years old, and again when she was twelve. Both times the girl blacked out; voices then filled her head and told her she was progressing well.

In 1950, when Andreasson was thirteen, came a more disturbing experience, according to her later account. After watching a moonlike object in the sky growing larger, Andreasson inexplicably found herself inside a strange white room where she was being examined by three small humanoids with big heads.

As Andreasson remembered the experience, they then placed her on a cushion in a spherical glass craft, which immediately plunged into a body of water. The vessel surfaced inside an icy tunnel, a section of which was lined with countless crystalline blocks. Inside the crystals were motionless human figures *(left),* embedded like flies in amber and dressed in the clothing of past eras.

Andreasson recalled that she then was transported away from this unnerving display into a dark area containing metal aircraft, from which she emerged into a place "like a forest of clear glass" where she experienced a sense of spiritual revelation. After returning the girl to Earth, the aliens admonished her to forget her travels. Another journey, however, still awaited her *(overleaf).*

A Terrifying Trip Past Bug-Eyed Beasts

According to her recollection under hypnosis, Betty Andreasson's second voyage to another world took place when she was thirty years old. Andreasson was living in South Ashburnham, Massachusetts, with her parents and her seven children, while her husband was in a local hospital recovering from an automobile accident. On the evening of January 25, 1967, the lights in the house flickered and went out. As the rest of her family fell into a trance, Andreasson said, she watched as five aliens entered her home by passing right through a closed door.

The strange visitors persuaded her to enter their vessel, where she was once again examined. She was then enclosed in a transparent canopy and flown to another world. There she and two extraterrestrials wearing black hoods were carried along a moving track through a long stone [illegible].

At the corridor's end, according to Andreasson, they passed through a silvery mirror into a region with a red atmosphere. Here, lemurlike creatures *(left)* crawled over square buildings at each side of the track. With huge eyes attached to stalks protruding from their headless necks, the beasts terrified Andreasson with their stares but let her pass unmolested.

Andreasson and the aliens are said to have then gone through a circular barrier into a vast underground realm suffused with a green light that revealed misty seas, lush vegetation, and a distant city. The wondrous voyage Andreasson later recalled was now almost at an end. Some four hours after she had left it, Andreasson found herself back at home, where she and her still-dazed family went to their beds. Andreasson retained only the foggiest memory of the encounter until visiting the hypnotist ten years later.

CHAPTER 1

In the Grip of the Strangers

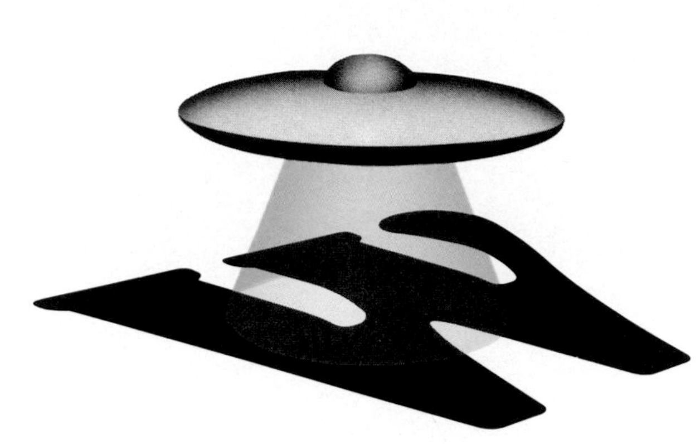

hen he finally talked to a doctor about his problem in 1988, Jim Weiner was at the end of his rope. For years the Boston man's sleep had been plagued by strange and terrifying dreams about creatures not of this world. Worse yet, he was often jerked out of slumber by a feeling that he was in the throes of a struggle against an unearthly force he could not see.

Unable to bear the strain any longer, Weiner told his troubles to a physician who had been treating him for epilepsy stemming from an accident ten years earlier. Weiner mentioned that the scenes in his nightmares reminded him of things he had read in a book about a man who claimed to have been abducted by a UFO. The doctor suggested that he attend a conference on UFOs soon to be held in a Boston suburb. Weiner took the advice, and at the conference he introduced himself to Raymond Fowler, a UFO investigator. As they talked, Weiner began to tell Fowler about an incident from his past that riveted the investigator's attention.

In August 1976 Jim Weiner, his twin brother Jack, and their friends Chuck Rak and Charlie Foltz had taken a camping and canoe trip to northern Maine. The outing began as a fun-filled adventure, including a stirring flight in a pontoon plane to the deep pine woods of the remote Allagash River country. But there, Jim related, something happened that cast a pall over the rest of the trip.

The men, he said, were out in their canoe on the night of August 26 when Chuck, in the rear position, suddenly got the feeling that he was being watched. He looked back over his shoulder and saw a large ball of bright light hovering noiselessly several hundred feet above the water. "That's a hell of a case of swamp gas," he exclaimed.

The other three turned to look; all were transfixed by the strange sight. They ceased paddling, and the canoe slowed to a stop. The sphere was as big as a two-story house, Jim Weiner later reckoned, and it seemed to have a liquid surface over which changing colors of light—red, green, and yellow—pulsated and swirled as it floated silently above the trees about two or three hundred yards away.

Charlie Foltz, who had been in the navy, began signaling to the object with a powerful flashlight the group had brought along, flashing the Morse code for SOS. The ball of light instantly began moving toward them, and suddenly what had been a mysterious sight to wonder at became a terrifying apparition. Three of the four started paddling desperately toward shore, the two in the middle, who had no paddles, slashing at the water with their bare hands. Chuck Rak alone remained curious and unconcerned, even when a cone-shaped beam of light shot out of the bottom of the UFO and traveled across the water toward them. "Swamp gas doesn't have beams!" yelled Charlie.

As the canoe neared shore, Jack Weiner thought, "This is it! We'll never get away." Then, abruptly, Chuck Rak was sitting serenely in the beached canoe and the others were standing on the shore, quietly watching as the glowing sphere seemed to curl in on itself and accelerate out of sight at dazzling speed. The men felt dazed; they were unable to move or speak for a few minutes. As the feeling wore off, Chuck got out of the canoe, and the group slowly went back to their campsite.

What the men saw there shocked them, even in their befuddled state. Before going out in the canoe, they had placed several huge logs on their campfire—enough, they knew, to burn brightly for two to three hours and help them find their way back in the inky night. Now, after they had been gone for what seemed no more than fifteen or twenty minutes, only embers and ashes remained of the fire. None of the men had checked his watch when the UFO appeared, but the evidence of the fire was unmistakable—and deeply disturbing. Someone or something that night had stolen two or three hours out of their lives.

For the rest of the trip, the men just went through the motions, scarcely remembering anything they did. And in the years to follow, the mystery of their missing time would remain unsolved. But not forgotten. After a number of years all four began to have the kinds of frightening dreams Jim Weiner told his doctor about, and all felt that their troubled sleep had something to do with that night in Maine.

Raymond Fowler, struck by Jim's mention of the missing time, soon got in touch with the other three men and by January 1989 had arranged for all four to undergo hypnotic regression, a technique by which a subject is encouraged, under hypnosis, to relive traumatic past experiences that are otherwise inaccessible to the conscious mind.

The story that the Allagash campers told while hypnotized was an extraordinary one. The men revealed that their memory of paddling toward shore in a state of extreme agitation, then suddenly standing calmly on the shore, had glossed over a long and terrifying ordeal. The beam of light they had tried to escape had actually caught them within its circle and drawn them up into the spacecraft. There they were confronted with alien creatures, who communicated instructions to them telepathically. As Jack Weiner related in halting speech while under hypnosis,

Four who claim they were kidnapped by aliens—from left to right, Charlie Foltz, Jack Weiner, Jim Weiner, and Chuck Rak—are photographed together in September 1990, fourteen years after allegedly being abducted while canoeing on Maine's Allagash River and two months after undergoing an exhaustive investigation of the incident. The affair is the first purported abduction described in detail by as many as four adult witnesses.

According to Chuck Rak, the beam of light in this picture he drew years later was the last thing he saw before the onset of a mysterious time gap. Charlie Foltz and Jim and Jack Weiner started paddling frantically when the light pierced the darkness and swept toward their canoe. Only Rak, seated in the rear, remained calm. "I felt no fear," he revealed under hypnosis, "only curiosity." In no time, the brilliant shaft caught up to the four men. Rak said that on looking up he realized that it was not a mere ray of light but was more like a tunnel and that he was being drawn up away from his companions, whose voices grew fainter as he ascended. At the end of the tunnel, he said, he passed through a barrier and found himself inside a circular room, where the other men soon joined him.

In the sketch at right, Charlie Foltz shows himself lying naked beneath some kind of scanning device while Jim and Jack Weiner and Chuck Rak (the figure indicated by a broken line at far left in the picture) wait their turn sitting on a bench along the wall. Rak recalled being probed with the apparatus that he rendered in the sketch below. "It's like they use in the dentist's office," Rak said of the alien instrument, "something with a long end, and it's got joints and it can fold and move."

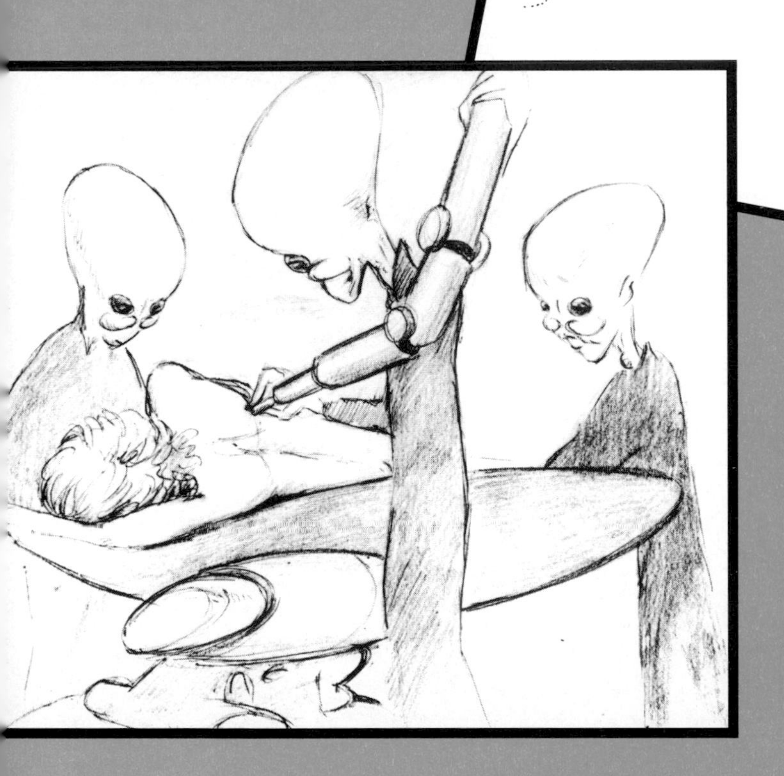

In Rak's drawing of the ordeal's conclusion (right), two aliens reposition him in the canoe. He said the four men were returned to the boat the same way they were taken: After they put on their clothes, they were brought into a circular room whose walls were ringed by a handful of narrow blue-gray tubes. An alien manipulated some instruments and a portal opened, causing bright, blinding light to pour into the chamber. As well as Rak can remember the incident, the men then walked to the edge of the portal and were gently floated down in a sitting position about sixty feet to the surface of the water.

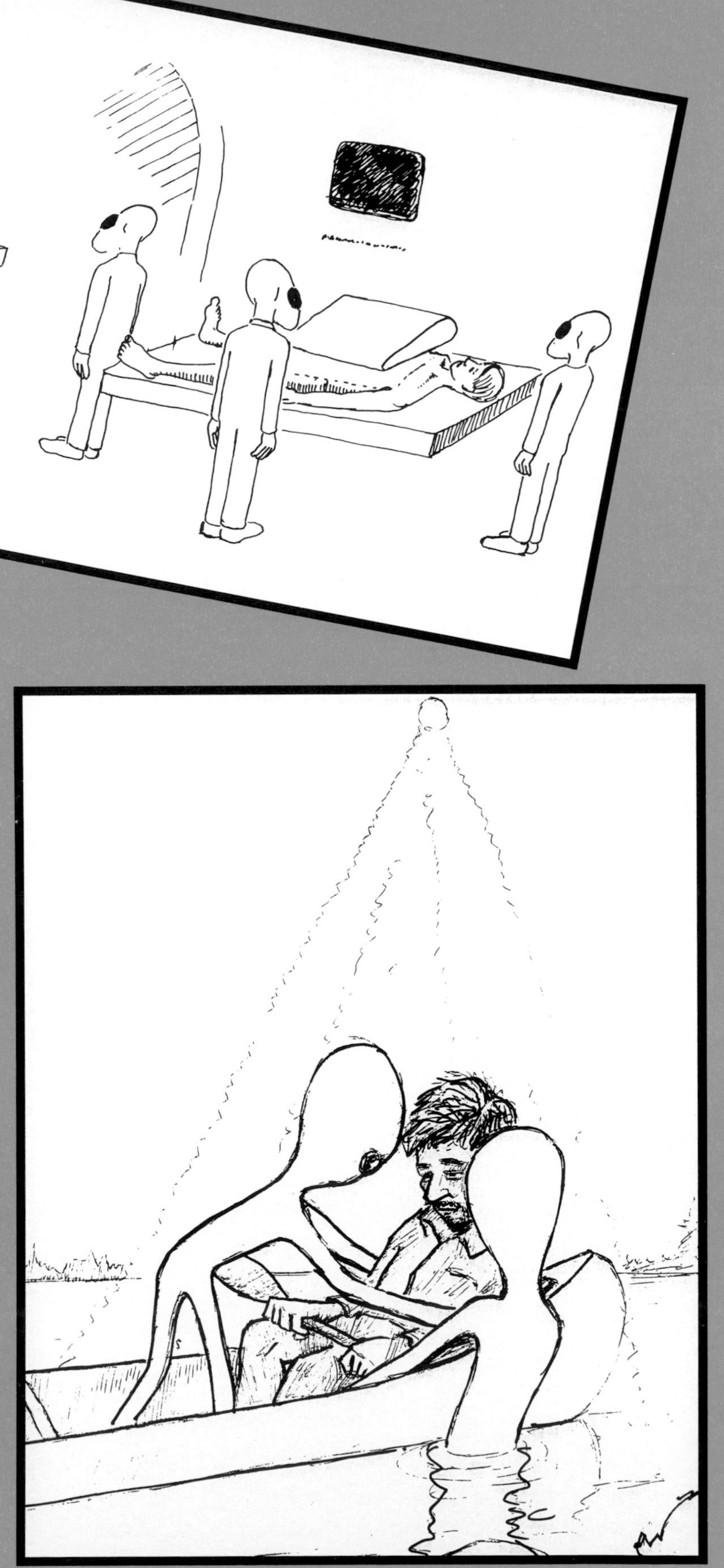

Have You Been Abducted? —The Signs to Look For

Abduction investigators say these telltale symptoms are common to people who purportedly have been kidnapped by aliens. But the same symptoms can indicate earthly problems requiring medical attention.

- Missing time. A memory gap, say believers, could mean a person has been abducted and was compelled by alien captors to forget the experience.

- Confused memory. Alleged victims often struggle to make sense of fragmentary recollections of mysterious lights, odd beings, invasive medical procedures, and other related phenomena.

- Irrational terror. A panicky fear is said to overcome abductees whenever they approach a particular location, see a helicopter hovering overhead, or find themselves in other situations with similarities to the repressed trauma. And they may react with anxiety to movies, magazine articles, and books that deal with UFOs and alien encounters.

- Nighttime disorders. Supposed abductees may have trouble falling asleep and then doze only lightly. Many dream of spaceships or bizarre creatures with oversize eyes, and some inexplicably wake up at the same moment night after night. In the morning, they feel disoriented and have short bouts of dizziness, numbness, tingling, and paralysis.

- Bleeding. Upon waking, some find blood on their pillows—an aftereffect, they believe, of the surgical implantation or removal of alien tracking devices in the nose or ears.

- Physical damage. Abductees also discover puzzling marks on their bodies—pinpricks, puncture wounds, scrapes, straight-line scars, small craterlike depressions, and bruises that they say are evidence of the physical examinations they endured.

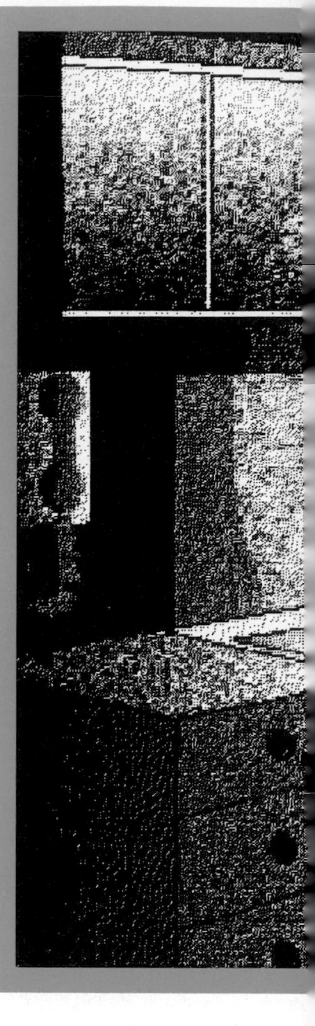

"They're saying things . . . with their eyes . . . in my head. They're saying, 'Don't be afraid. . . . We won't harm you. . . . Do what we say.' "

All four were made to undress, lie on a table, and undergo a painful and frightening physical examination. Then, the next thing any of them knew, they were back on the Allagash—Chuck in the canoe, the others on shore—and the UFO was departing the scene.

If the four men had told such a story twenty-five years earlier, it would probably not have been taken seriously. Even UFO enthusiasts, people who believe unquestioningly in the existence of visitors from other worlds, would most likely have dismissed it as a fantastic dream. But by 1989, what the Allagash four revealed under hypnosis struck several familiar chords. In particular, the phenomenon of missing time—an inexplicable gap in a person's life, typically of an hour or two, sometimes longer—had become well established as a common element in similar reports from all parts of the United States, as well as from Europe, Asia, Australia, and Latin America. The theme linking the stories was the narrators' conviction, sometimes after hypnotic recall, that they had been abducted by aliens.

There was nothing new about the idea of alien contact in itself. Ever since the postwar UFO wave was touched off by the first reports of flying saucers in 1947, there had been speculation about the creatures that might be riding inside the supposed spaceships, and in the decade that followed, several individuals would claim to have been contacted by these extraterrestrial voyagers.

The best known of the "contactees," as the 1950s observers were later named, was George Adamski, a Polish-

Hovering globes of light like the one pictured here by Allagash abductee Jack Weiner, a computer artist, are said to appear at times in the lives of abduction victims, perhaps performing a monitoring function. Weiner remembers this one entering his home when he was fifteen. The orb floated from the kitchen to his bedroom before wafting out a window and vanishing.

born odd-job man from California who wrote books describing his encounters with the visitors. The beings he claims to have seen were generally tall, blond, and stately. They had come to Earth from their home on Venus, he said, to deliver a warning about the dangers of nuclear weapons testing. Adamski asserted that the aliens took him on flying-saucer rides to the far side of the Moon, where he saw great cities—a statement that considerably deflated his already tenuous credibility, particularly when, in 1959, Soviet satellite photographs showed the far side to be the barren wasteland astronomers had always assumed it was.

The derision rubbed off on other contactees of the time, who, like Adamski, tended to encounter handsome, human-looking beings prone to rambling philosophical discourses. Their loquacious concern for the Earth's well-being won these first-generation aliens the appellation "Space Brothers" among early observers of the UFO phenomenon. For some years, the scant credence accorded Space Brothers stories effectively discouraged serious investigation of any purported contacts with aliens, even by flying-saucer enthusiasts. Attention focused instead on attempts—most notably by the U.S. Air Force's Project Blue Book—to verify any of the continuing stream of reported UFO sightings, and on the hardware and technology of the new field of investigation called ufology.

What changed the situation was the now-famous story of Barney and Betty Hill. The 1966 publication of *The Interrupted Journey,* a book recounting the New Hampshire couple's alleged abduction by aliens five years earlier, and later the broadcast of the widely seen television movie based on the book, brought to public notice a fascinating sequence of events unlike anything to be found at that time in the growing literature of extraterrestrial contact. The Hills' account of being snatched from their car by a UFO, subjected to terrifying physical examinations, and returned to the car with no conscious memory of the ordeal had most of the characteristics that investigators would later class as typical of abduction stories. Once the Hills' saga hit the mass media, the era of the alien abduction experience—or, more properly, of abduction reporting—was under way.

The elements in the Hill story, and in a similar tale told by a young Brazilian by the name of Antonio Villas Boas, who claimed to have been abducted in 1957, soon started to show up in a growing number of other accounts reaching UFO investigators and, sometimes, the media. A number of the stories detailed events that purportedly had happened well before the Hills' and Villas Boas's experiences but that had been kept secret by the abductees. Only after observing the publicity surrounding the Hills' ordeal did the individuals involved feel confident enough to admit publicly their own experiences. Some of the victims, like Villas Boas, had conscious recall of what had happened to them; others, like the Hills and the Allagash four, were aware only of a sense of missing time until much later, when their traumatic memories of abduction were retrieved by hypnosis.

Some of the reports received massive publicity. A triple abduction in 1976 threw the limelight on three Liberty, Kentucky, women, whose drive home late one night was interrupted in a terrifying fashion. Louise Smith and Elaine Thomas had taken their friend Mona Stafford out to dinner in nearby Lancaster to celebrate her thirty-sixth birthday. Near the town of Stanford, they said, they saw a reddish, disk-shaped UFO with a white, glowing dome. The craft followed their car, illuminating it with a blue light. Louise Smith, who was driving, lost control of the vehicle, which accelerated to eighty-five miles per hour, and the women felt burning sensations in their eyes. They experienced hallucinations of being dragged backward in the car along a bumpy road, and then they saw a wide, brilliantly lit highway that was unfamiliar to them, although they knew the area well. The women eventually reentered normal reality at a point eight miles from the original encounter site and drove home without further incident, but with a mysterious time gap of more than an hour.

The women all subsequently suffered severe anxiety symptoms and drastic weight loss. Persuaded to undergo hypnotic regression, they each related a similar story of abduction by four-foot-tall aliens, followed by painful physical examinations. In the weeks and months after the event, all of them claimed to be experiencing various psychic phenomena, which they associated with the abduction. After the incident, Elaine Thomas suffered ill health and eventually died three years later at the age of fifty-two. Oddly enough, of the three victims she alone had claimed to have experienced beneficial personality changes, asserting that the alien encounter had somehow made her more self-confident and outgoing.

The victims of one alleged abduction labored under the mystery of what had happened to them for more than twenty years before an answer was finally dredged up from deep within the psyche of one of them. The story began in March 1953, when two women in their early twenties were sharing a cabin in Tujunga Canyon, just north of Los Angeles. Sara Shaw—a pseudonym chosen to protect her privacy—said that she woke up suddenly one night, aware that something was eerily different in the house around her. There was an unnatural hush; something had deadened the usual night noises. Then, she said, as she

Dissimilar in age, personal circumstances, and place of residence, the people pictured here—from far left, Kentucky friends Louise Smith, Elaine Thomas, and Mona Stafford; Arizona logger Travis Walton; and Connecticut homemaker Betty Andreasson—all stated that they had experienced abduction by alien beings.

lay between sleep and waking, she saw that a light was shining in through the picture window at the head of the bed. It had a curious bluish tinge, and it seemed to be sweeping back and forth.

Shaw got up and looked out into the yard but could not make out where the light was coming from. Fearing intruders, she awakened her companion, even though a glance at the clock showed her that it was 2:00 a.m. Sensing Shaw's alarm, the other woman leaped out of bed and moved toward the closet to get her robe.

uddenly, for each of the two women, the real world fell away. They sensed giddiness, panic, then blankness. And although the disorientation seemed only momentary, when they came to themselves again, the peculiar light was gone and the clock was showing 4:20 a.m. The air in the cabin seemed stuffy, and they found they were having difficulty breathing. Terrified now, the women ran out to their car and drove to the home of a relative. When they summoned up the courage to return to the cabin a couple of days later, they found it undisturbed. They moved back in and resumed the life that the strange light had disrupted.

But the memory of that night continued to trouble Sara Shaw over the years to come. It was still sufficiently strong that after watching a television documentary on UFOs in 1975, she decided to seek out someone with whom she could talk over the incident. When the UFO investigators she reached suggested hypnosis, she agreed.

Under hypnosis Sara Shaw related that she and her companion had not been alone in the cabin that night, as they had always supposed. In the two-hour-and-twenty-minute gap of lost time, their home had been invaded by a number of tall, thin aliens who had led her and her companion to a spacecraft, which turned out to be the source of the blue light the two women had seen earlier. The craft was circular and girded by a platform that Shaw compared to the rings of Saturn.

Inside its domed interior, the women were stripped naked and forced onto tables, where they were physically examined by their captors. Shaw said that before being returned to their cabin the two women were given instructions to forget that the encounter had ever taken place. There the matter uneasily rested, until Sara Shaw's initiative in bringing the mystery to light gave her story a prominent place in the rolls of alien abduction reports.

More than any other abduction claim made up to that time, a 1975 incident in Arizona at first glance seemed to offer objectively verified proof of the reality of the abduction phenomenon. A logger by the name of Travis Walton was said to have been assaulted by a beam from a spaceship while no fewer than six fellow members of his work crew witnessed the event. Not seen again for five days, Walton returned with tales of large-eyed aliens who had "beamed" him into the spaceship, examined him, then conducted him out of the craft into a hangarlike room. He was then moved to another chamber but had little memory of what happened after that.

Drawn under hypnosis, these hairless, almond-eyed humanoids bear an uncanny resemblance not only to each other but also to creatures described by hundreds of other alleged abductees. These were reportedly seen by Betty Andreasson during a 1967 abduction (left), Jack Weiner in the 1976 Allagash case (center), and Kathie Davis, who claimed that aliens removed a fetus from her womb in 1978 (below).

Some serious doubts, however, were soon cast on the reliability of the Walton story. Philip J. Klass, a longtime skeptic on the subject of flying-saucer phenomena, discovered that the alleged abductee was known to be a UFO enthusiast and that at the time of the incident the team of which he was a member had fallen behind on a logging contract and faced possible financial penalties. But the contract contained a clause that forgave noncompletion if it was caused by an act of God. Klass suggested that the loggers, finding themselves in a financial bind, could have dreamed up the notion of faking an abduction, reasoning that such a terrifying event at the work site would qualify as an act of God and get them off the hook. Furthermore, Klass pointed out, the incident occurred only sixteen days after the broadcast of the television version of *The Interrupted Journey,* when the theme of abductions was very much in the public mind. The much-sought-after ironclad proof that people were being abducted would not be found in the Travis Walton case.

The period from the mid-1970s to the early 1980s proved to be banner years for abduction reports. Twenty-five cases emerged in 1975 alone. In 1979, twenty-seven cases came to light, and more than forty were logged in each of the two following years. To deal with the growing number of cases that needed investigating, several new volunteer ufologists' organizations came into being. The J. Allen Hynek Center for UFO Studies (CUFOS), the Fund for UFO Research (FUFOR), and the Mutual UFO Network (MUFON) joined the ranks of such venerable groups as the Aerial Phenomena Research Organization (APRO) and the National Investigations Committee on Aerial Phenomena (NICAP), which were both founded during the 1950s.

In 1987 an abduction story shot to the top of the best-seller lists. In *Communion,* popular horror-story writer Whitley Strieber recounted experiences that he said had happened to him in his isolated cabin in upstate New York. According to the book, he was abducted from his bed by an alien while his wife lay sleeping at his side. Taken aboard a craft, he encountered other aliens of differing shapes and sizes; one, whom he took to be female, inserted a needle-like probe into his brain. His account of the horrific experience, recalled with the aid of hypnotic regression, would make *Communion* an even bigger seller than his earlier, fictional works, *The Wolfen* and *The Hunger.*

By the time of the book's publication, so many Americans were claiming to have experienced abduction that the

Not all aliens fit the same description. The creature on the left, reportedly seen by Fortunato Zanfretta in 1978, stood ten feet tall. The one on the right, which Carl Higdon claimed abducted him in 1974, had arms that ended in strange devices instead of hands and "hair that stood like straw."

event was approaching the level of a psychosocial phenomenon. Support groups, in which people who claim to have been abducted shared therapeutic discussions about their problems, sprang up in several parts of the country. There was even an annual conference for abductees, first convened in 1980 at the University of Wyoming by a longtime UFO investigator, Leo Sprinkle, who often served as a consultant to APRO. In 1991, 137 people attended the conference.

To impose some order on the welter of extraterrestrial incidents being reported, one noted researcher, J. Allen Hynek, had developed in the early 1970s a hierarchy of types of contacts. His system began with three low-ranking categories of distant UFO sightings—the lowest being lights in the night sky, followed by disks seen in daylight and sightings backed by radar readings. Of greater moment were his three groupings of more direct contacts, which he named close encounters.

A close encounter of the first kind was a sighting of a UFO from no farther away than 500 feet—but without any interaction between the UFO and the observer or the environment. Discernible physical effects attributable to UFOs, such as patterns of crushed or burned vegetation in cultivated fields, or shutdowns of electrical systems, were close encounters of the second kind. A close encounter of the third kind, in this typology, was one in which alien beings were seen in or near a craft.

Hynek, the sober and scholarly founder of CUFOS, went no further with his ranking system, wishing to avoid being lumped with the more or less disreputable "contactees" of the 1950s and their dubious Space Brothers. But the growing numbers of abduction claims from the mid-1970s onward made it clear that the close encounters of the third kind category—abbreviated CEIII by ufologists—failed to accommodate an increasingly important phenomenon. A CEIV has since been added: abduction of humans by aliens.

Although the bulk of abduction reports still came from the United States, similar stories were also being uncovered in other countries where organized groups of UFO observers were active, notably in Argentina, Brazil, the United Kingdom, and Australia. One account even told of a woman parachutist who vanished during free fall, only to reappear three days later claiming to have been intercepted by a flying saucer in midair.

There were by the late 1980s enough self-proclaimed abductees for investigators to attempt demographic analyses of their backgrounds and personalities. The victims were shown to have come from all walks of life, although their overall educational level was slightly higher than the general average. About two-thirds were male, with a slight preponderance of individuals involved in work that took them out-of-doors on their own at night—police officers, farmers, truckers, traveling salespeople. Abduction was for the most part a solo experience.

The aliens' only obvious predilection seemed to be for young adults. A disproportionate number of victims were in their twenties and thirties; a seventy-seven-year-old Englishman claimed to have been abducted in August 1983 but then released after only a cursory examination by his captors, apparently because he did not fill the bill.

There was much less uniformity when it came to reporting what the aliens themselves looked like. Abductees described almost every conceivable kind of creature, from giants to dwarfs, from the humanoid to the bizarre. Figures reminiscent of the Space Brothers appeared in some reports; a Brazilian truckdriver who said he was taken while changing a flat tire one night described his captors as two short-haired blond men and a long-haired blond woman, all wearing silvery clothing. The creatures in a well-publicized Italian case, however, were less fetching. A twenty-six-year-old security guard, Fortunato Zanfretta, was discovered by colleagues lying in a meadow in a confused state after radioing his base station for help. Under hypnosis, he subsequently told of being abducted by ten-foot-tall monsters with green skin, yellow triangles for eyes, red veins on their heads, and pointed ears.

The creatures that carried off two fishermen near Pascagoula, Mississippi, had eyeless faces and clawlike hands; their gray skin resembled wrappings or bandages, giving them a mummified look. A twenty-one-year-old who claimed he was snatched from his car in 1975 saw beings about four and a half feet tall with mushroom-shaped heads and no visible mouths or ears. Oddest of all, perhaps, was the creature who in 1974 reportedly took control of a Wyoming man named Carl Higdon while he was out hunting. The being, who said his name was Ausso, was tall and unprepossessing, a chinless, bow-legged specimen with only a few teeth, sparse sprigs of hair, and weird implements in place of hands.

These, though, were the oddities. From a majority of the reports, and particularly from the better-documented ones, a more consistent picture emerged. It was one of humanoids, somewhat shorter than the average Earthling, with large heads, large eyes, and spindly frames. Typical were the beings who appeared to Betty Andreasson, the New England mother of seven whose reported series of abductions, stretching over almost four decades, furnished the material for three separate books. Recalling a 1967 abduction, she described creatures between three and a half and four and a half feet tall. Their faces had an inverted pear shape and were—by human standards—disproportionately large for their bodies. They had large, catlike eyes extending around to the side of the head. Their skin color was gray, and they wore close-fitting uniforms.

Similar descriptions regularly cropped up in other reports. Under hypnosis, Sara Shaw haltingly described one of her abductors: "His head is—is elongated. It's not even egg-shaped. It's like an egg, but one that isn't really wider at the top or the bottom. . . . it's oval; it's oblong. . . . If there's any hair, it's underneath this skintight—it's like a ski mask, but it's almost part of the skin. It's as if he were sprayed with a can of flocking; as if he were wearing a coat of flocking. . . . He has eyes that are like a ski mask—just like a ski mask, with two openings for the eyes, but I see only what we would call flesh color. I don't see eyeballs, or whites, or lashes. It's almost like . . . flesh-colored membrane."

There was less variety in descriptions of the crafts in which the aliens traveled. As might have been expected, they were usually saucer shaped, although reports of oval or oblong ships were also common. Inside, the principal chamber was usually round and domed. For psychologically minded skeptics, the absence of corners in the descriptions suggested that the victims were subconsciously recalling images of the womb.

Given the sheer number of claimed abductions, more than 300 by 1985, wide variety in the stories might have been expected, but in fact the victims' reports tended to follow a regular pattern. Although they often differed a great deal in details, there was a sequence of abduction, examination in a brightly lit room, and return to the familiar world that appeared again and again.

When a folklorist, Thomas E. Bullard, analyzed the 300-plus abduction reports, he was able to distinguish eight typical stages in the process. The first, of course, was *capture.* Next came *examination;* followed by *conference,* a phase in which the aliens held a telepathic conversation with the abductee; a *tour* of the spacecraft; an *otherworldly journey,* in which the abductee was sometimes transported over great distances; then the rarest event of the sequence, a religious experience labeled by Bullard a *theophany.* A *return* to the normal world marked the conclusion of the abduction itself. Bullard's last stage was the often damaging *aftermath,* in which the victims continued to be troubled, sometimes for long periods following the abduction experience. Although only one of Bullard's cases included all eight stages, virtually all featured several of them, usually in the same order of occurrence.

In the first of the stages, capture, light played a crucial part. The abductee was often first alerted to the presence of something abnormal by the sight of a strange illumination; in other cases a beam of light was instrumental in drawing the captive up into the alien craft. Many victims described a sensation of floating as they moved toward the UFO, although the feeling may have been deceptive: Barney Hill thought he had floated into the spacecraft, but his wife, who was watching, reported that he was actually walking, half-supported by his captors. One of the oddest features of the capture was that very few abductees seemed to retain any recollection—even under hypnosis—of the moment of entry into the craft, a circumstance that Bullard characterized as "doorway amnesia."

In the domed interior of the vessel, the central drama of the abduction experience was played out. Here, in 133 of Bullard's cases, the abductees underwent an often painful physical examination. This episode loomed so large in so many accounts, and was so consistently the first thing the victims were subjected to upon entering the spacecraft, that those who accept the reality of the abductees' claims have little doubt that the examination is the primary purpose of the entire abduction phenomenon. The aliens' long-term goals, they believe, are somehow inextricably connected with the detailed investigation of human biology.

Often the examination fell into separate parts. Initially the victim, laid full length upon a table, might be scanned by a device similar to an x-ray camera. Sara Shaw tried to

Exploring "Missing Time"

Once regarded only as a source of entertainment, hypnosis today is commonly employed in the fields of medicine, dentistry, psychotherapy, and even law enforcement, and it has become one of the primary tools of alien abduction investigators the world over.

Most of the time, experts use hypnosis simply as an investigative instrument, a means of uncovering what happened during the periods of missing time that torment many alleged abductees. Subjects are urged to relive and describe experiences that they have—on the orders of alien captors—repressed. But hypnosis can also be a therapeutic tool to relieve the anxiety and stress suffered by people who believe they have seen a UFO or who think they have been abducted by aliens.

Using hypnosis to uncover supposedly repressed memories is controversial: Skeptics contend that subjects may distort or fabricate information and that their recollections often reflect their own fantasies, fears, and desires more than they reveal evidence of alien beings. UFO investigators readily concede that this is possible. Nevertheless, they point out that the experiences related under hypnosis by most abduction victims are remarkably alike—much too similar, in fact, to justify dismissing the abduction phenomenon out of hand.

Hypnotist Tony Constantino (left) places Jack Weiner (right), one of the four men involved in the Allagash affair of 1976, into a trance on April 15, 1989, while veteran UFO investigator Raymond Fowler looks on.

describe the one she saw: "It's like the other half of the table. But, you know, upside down, sandwichlike. It's not scorching me or coming anywhere near me. It's just hanging over me."

After the scan, the aliens generally examined the abductee directly, sometimes using instruments to take samples of skin and other tissue, as in the Hill case, or even gouging out tiny scoops of flesh, leaving lasting indentations in the skin of their victims. New York artist and ufologist Budd Hopkins particularly noted this phenomenon, stating that the operation could leave scars of two different types: round, shallow scoops or long, thin, scalpel-like cuts.

The examination was often a terrifying ordeal. The victim usually felt a devastating sense of loss of control; he or she was completely at the mercy of the aliens, whose interest in the proceedings generally appeared to be as clinical and objective as that of human scientists working on laboratory animals—and often as devoid of warmth or consideration. A twenty-seven-year-old woman who relived an abduction experience under hypnotic regression said: "I was so scared. I knew they were so strong. I think I gave up. . . . I just laid there. . . . It was a terrible feeling."

In the Allagash case, where four victims at once were held in the spacecraft, the aliens used some covert form of behavior control to maintain order. Jim Weiner testified that while his twin brother Jack was being examined, the other two members of the party sat on a bench with a "dumb, expressionless look" on their faces, while he himself could only look straight ahead. Since the others independently reported under hypnosis the same vacant look on Jim's face, the investigators concluded that they all had been put in a state of suspended animation.

fter the rigors of the examination, Bullard noted, the atmosphere often lightened as the phase he described as the conference got under way. "The beings relax, slow down and warm to their captive," he wrote, "often taking him out of the examination room with its unhappy memories to another part of the ship. This change to a friendlier, more considerate atmosphere is striking, since the beings suddenly begin to treat their captive like a human being and even a guest, instead of like a guinea pig."

The aliens would now sometimes carry on an informal conversation—telepathically—with their victims. Occasionally their words, as reported by abductees under hypnosis, recalled the well-meaning platitudes of the Space Brothers. Betty Andreasson's abductors told her that they were "the watchers," caretakers of nature and natural forms. "They love mankind," she asserted. "They love the planet Earth and they have been caring for it and Man since Man's beginning. They watch the spirit in all things. . . . Man is destroying much of nature. They are curious about the emotions of mankind."

According to other abductees, their captors evinced a similar curiosity about earthly matters but seemed to be motivated more by self-interest. A fifty-three-year-old California businessman reported that his aliens were interested in humans because their own society, though technologically highly advanced, had lost qualities they still saw among humans. "What has happened with

Betty Andreasson made this painting to illustrate how the alien beings that visited her in 1967 passed through a closed wooden door. She said they moved "in a jerky motion"—vanishing and then reappearing, each time "leaving a vapory image behind."

them is they lost their identity, their individuality and their uniqueness," he claimed.

In some reports, the abductors offered more specific information, including hints as to their place of origin. Betty Hill was allowed to see a star map containing the home planet of her captors, but—as the being who showed it to her pointed out—the knowledge was of little use to her as she had no way of orienting it to Earth. Nonetheless she was subsequently able, under hypnosis, to sketch an approximation of what she saw.

An Ohio schoolteacher named Marjorie Fish was later to spend five years attempting to correlate the sketch with known star locations, eventually suggesting that the aliens' home planet might have been located in the neighborhood of Zeta 2 Reticuli, one-half of a twin-star system in a small constellation visible in the Southern Hemisphere, thirty-six light-years from Earth. Other astronomers were less convinced. Astrophysicist Carl Sagan subsequently used a computer to plot Fish's star positions and concluded that they bore little similarity to those in Hill's drawing.

Prophecies of future events occasionally delivered by the aliens have proved even less persuasive. A U.S. Air Force staff sergeant claimed that his captors informed him that they would return publicly within three years; they did not. The Pascagoula fishermen supposedly were similarly told to expect another visit from their captors before 1983, in which they would turn humankind from its destructive ways; the world is still waiting.

The failed predictions have provided powerful ammunition for skeptics. In rebuttal, believers in the reality of abductions point to a pattern of deceit discernible in the aliens' behavior—hiding their real motivation from their victims, for example. Believers assert that the false prophecies are a smoke screen having the twofold purpose of masking the abductors' actual designs and undermining the credibility of the abductees' claims to have seen them.

After the conference stage, Bullard reported, sixteen of the abductees covered in his study were given a tour of the craft. One such was Herbert Schirmer, a state trooper who reported sighting a UFO on a Nebraska highway in December 1967 and subsequently revealed an abduction experience under hypnosis. He claimed to have been escorted on what amounted to a technical round of inspection. In its course he learned that the craft was made of pure magnesium, and he was told that it drew electricity from power lines on Earth simply by pointing an antenna at them. He was left in the dark about why the craft might need electricity and what happened when it was out of range of Earth. The ship was propelled by "reversible" electromagnetic energy that created an inertia-free, gravity-free flight. Schirmer was shown a crystal-like circular object that he was told powered twin reactors—whether nuclear or some other kind was not made clear, nor was he told why they were needed if the ship ran on "electromagnetic energy." He also saw a large viewing screen that enabled the occupants to study their surroundings even in the dark. The on-board gear included a

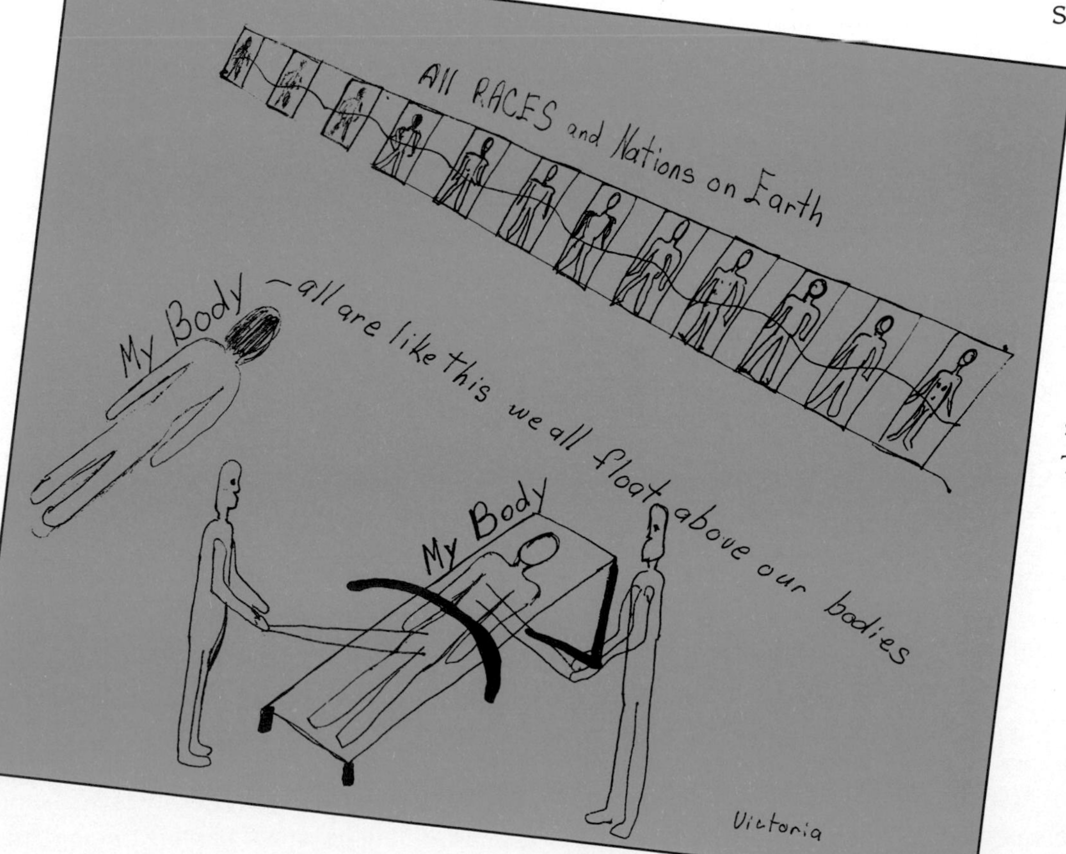

Asked to sketch an abduction she remembered under hypnosis, an artist in her fifties depicted herself floating above her own body while aliens remove her ova. Men and women from every corner of the world (top), she said, were subjected to similar operations.

number of smaller saucers, apparently used for scouting.

Accounts of such tours were normally less informative than investigators hoped, because the witnesses lacked the knowledge or even the words to explain what exactly they were seeing. The craft they described resembled for the most part standard movie images of spaceships, with engine rooms, control rooms featuring computers and star charts, and, sometimes, living quarters. Few included details as striking as the vivarium that Betty Andreasson reported having seen during one of her abductions. It seemed as she entered it to be a separate world of woods and water, where she saw a pond with fish swimming in it. Only when a door opened and light flooded in did it become apparent that this pastoral environment had somehow been created on board the spacecraft.

An even more spectacular appeal to the imagination was provided by Bullard's fifth stage, the otherworldly journey, a feature he found in fifty-four of the reports that he categorized. He used the term to describe all voyages, some of them merely terrestrial, made by abductees aboard their captors' craft. Among reported destinations were New York City, the pyramids of Egypt, a U.S. Navy destroyer, and a supposed UFO landing field at the North Pole.

The most interesting reports of otherworldly journeys, however, featured travel to entirely unearthly realms. Some witnesses described bare, lifeless scenes, with stunted trees rising against empty skylines. Other abductees saw lush landscapes rich in alien plant life. There were descriptions of futuristic cities with suspended roadways, and of factories manufacturing UFOs or crystals, apparently used for the production of energy.

Probably no accounts were more extraordinary than the descriptions Betty Andreasson gave of her travels with aliens. When hypnotized in 1977 and 1980, Andreasson recalled that she had been abducted from her home twice: in 1950, when she was thirteen, and again in 1967. Her purported experiences on these occasions have a decidedly religious, mystical cast to them. She told of passing through water in the alien craft and emerging into a realm inhabited

Recalling her otherworldly journey of 1950, Betty Andreasson painted herself lying on a disk of jellylike material and wearing an alien device to hold her tongue in place during rapid acceleration.

by frightening, unearthly life forms and of a dreamlike world bathed in light and containing floating crystal clusters and a pyramid. At the end of one journey was a creature that could have emerged from the pages of the Book of Revelation. She saw a gigantic bird, resembling an eagle with a long neck, that as she watched was, like the phoenix of ancient legend, consumed by fire—except that from the ashes crawled a fat gray worm.

During one hypnotic regression devoted to Andreasson's 1950 abduction, her recollections provided a memorable example of the mystical experience that Bullard called a theophany, the sixth of the stages he listed, although it appeared in no more than 6 of the 300-plus cases he logged. On this occasion Andreasson spoke of passing through a subsurface tunnel to a glass wall with a great door. Beyond this door, she was told, was a being called simply the One; Andreasson, a religious woman, seemingly identified the being with God. She then saw a twin image of herself, perhaps representing her spirit, passing through the

May 12, 1980
WALLS ROLLED UPWARD AND WERE OF GLASS.
LAYERS OF THICK CLEAR GLASS.
ME.
I CAME OUT OF ME, AND A SHELL OF ME (?) STOOD MOTIONLESS
HUGE CLEAR GLASS BALL

The climax of Andreasson's fantastic voyage was her arrival before a many-layered glass passageway, shown here in a drawing she made after being hypnotized. "Now you shall enter the Great Door and see the glory of the One," said her guide, a three-foot-tall humanoid with a large, pear-shaped head and dark, unblinking eyes. Andreasson claims she came out of her body and floated away from the crystalline sphere (below, right) that had transported her. She then stepped through the portal. What she saw on the other side, however, remains untold: "Words," she contends, "cannot explain it."

door, to reemerge later with a look of radiant joy on its face.

At the time, her hypnotist repeatedly asked her to describe what she saw behind the door, but she insisted that she could not do so. A further attempt to unlock her memory eight years later also failed, even though the hypnotist, in an attempt to lessen the emotional content of the experience, suggested that she observe it as though it were projected on a television screen. "There's a bright light coming out of the television," Andreasson responded. "This is weird! There's rays of light, bright white light, just like they've got a spotlight coming out of the television. . . . It's too bright! It's hurting my eyes!" Recognizing that she was in pain, the hypnotist brought the session to a close, and Andreasson reported headaches and aching eyes for several days afterward.

In comparison with the wonders of the otherworldly journey, Bullard noted a sense of anticlimax in descriptions of the return, the seventh of his categories and the last extraterrestrial stage of the abduction experience. "This episode gets slighted in many reports," he wrote, "because it simply reverses capture without adding much that is new to the story."

A notable exception, however, was the testimony under hypnosis of Chuck Rak, one of the Allagash four. Rak reported that when it came time to return to Earth, he and his companions were led to the wall of the spacecraft where, in the manner of movie or television science fiction, a portal somehow appeared. "It's almost like a place in the wall where something happens," said Rak. "And it's like we're penetrating a membrane."

The Allagash exception notwithstanding, the usual experience is for the abductors to bid farewell to their captive, perhaps enjoining secrecy or inducing memory loss. Then the victim is returned to their normal earthly world—although after the abduction experience, normality may seem a dubious state to the returnee.

Such, at any rate, was the pattern for those who did return. But the annals of UFO abductions include a few cases in which the victim was never seen again. One such case concerned a twenty-year-old Australian pilot by the name of Frederick Valentich. At 6:19 p.m. on Saturday, October 21, 1978, Valentich took off from a Melbourne airport in a single-engine Cessna 182. His destination was King Island, 150 miles to the south across the Bass Strait.

orty-seven minutes into the flight, Valentich radioed air-traffic control to report what he took to be another aircraft in his vicinity. He described seeing four bright lights. The tower replied that they had nothing out of the ordinary on radar. Valentich insisted that he could see a craft, which he described as elongated. "It's coming for me right now," he reported. "It's got green light," he said, and "is sort of metallic; it's all shiny, the outside."

Moments later he reported that the unidentified object was circling above him. "It seems to me that he's playing some sort of game; he's flying over me two, three times at speeds I could not identify."

Just after 7:10, Valentich reported that the craft was now apparently holding position above him, even though he was circling to try to evade it. A couple of minutes later he radioed that his engine was surging and cutting out. His last transmission was: "That strange aircraft is hovering on top of me again. . . . It is hovering and it's not an aircraft." The radio operators then heard a scraping, metallic noise.

Neither Valentich nor his plane was ever seen again, although the area of the disappearance was thoroughly searched. Reports were received of UFO sightings in the Bass Strait at the time, but suggestions that Valentich and his craft were abducted must, like many less dire UFO stories, inevitably remain speculative.

The lack of confirming evidence in the Valentich case reflects the central problem of the abduction accounts themselves: an absence of hard evidence. For the most part investigators trying to assess the objective reality of the experience have nothing to go on but the abductee's word as to what happened.

A UFO Buff Takes Leave of This World

Even though Granger Taylor never finished the eighth grade, people who knew him insist he was hardly an ordinary grade-school dropout. Some even call him a genius—a young mechanic so gifted he could single-handedly bring rusty bulldozers and long-forsaken tractors back to life and return decades-old cars to showroom splendor.

Taylor worked as a repairman's helper for about a year after quitting the classroom, then began tinkering for himself on his parents' wooded property in Duncan, British Columbia, a lake resort at the southeastern end of Vancouver Island. In no time, the lot took on the appearance of a verdant junkyard: Vintage cars sat beneath stands of fir trees, and welding equipment and scavenged motors lay scattered among the bushes.

Taylor's accomplishments, however, more than made up for the eyesore. When barely into his teens, Taylor reclaimed an abandoned steam locomotive and restored its vigor. At fourteen he built a novel one-cylinder car that was displayed—along with the train engine—at the Duncan Forest Museum. A collector reportedly shelled out $20,000 for a replica World War II fighter plane the youth had built.

Despite such successes, Taylor apparently lost interest in earthly modes of transportation in the late 1970s, and he turned to a different class of vehicles—UFOs. Curious about how flying saucers might be propelled, Taylor read book after book dealing with alien beings and spacecraft, and he pondered the question for days at a time in the flying-saucer-shaped house pictured above, which he constructed from two satellite dishes and outfitted with a television set, a couch, and a wood-burning stove.

According to longtime friend Bob Nielson, Taylor was in his residential "spaceship" in October 1980 when he received communications from alien beings. "He couldn't see them," Nielson says. "They were just talking to him and to his mind." Taylor asked the beings how their spacecraft were powered, but the aliens would not tell him the answer. Instead, they invited him to travel with them through the solar system, and he accepted.

About a month later, Taylor tacked a note to the door of his father's bedroom. "Dear Mother and Father," it read, "I have gone away to walk aboard an alien ship, as recurring dreams assured a forty-two-month interstellar voyage to explore the vast universe, then return. I am leaving behind all my possessions to you as I will no longer require the use of any."

That night, a violent storm swept ashore from the Pacific Ocean. Hurricane-force winds lashed Vancouver Island and downed electric lines, and the city of Duncan was plunged into darkness. When the sun rose the next morning, Granger Taylor and his pink pickup were gone. Six years later, the truck was found on a mountain near Duncan, apparently blown to bits in what must have been a massive explosion. But Taylor's body was never found.

One phase of Bullard's abduction sequence, however, did have obvious manifestations. That was the phenomenon he labeled the aftermath. It was undeniable that in many cases individuals reporting abductions presented physical and psychological symptoms suggestive of highly stressful experiences. Sometimes these effects rapidly wore away, but in other cases they left lasting scars.

Some aftereffects were physical. Many victims reported persistent nausea and headaches as well as heightened sensitivity to touch. Abductees might show burn marks; in some cases they developed rashes. A claim by Antonio Villas Boas that a bruise took abnormally long to heal after his abduction suggested to some that his immune system might have been affected. Barney Hill developed a circle of warts in the genital area; in his testimony under hypnosis he had claimed that his abductors had placed a cuplike device over his groin.

Many victims mentioned suffering burning or bloodshot eyes. The three women abducted in the 1976 Kentucky case all subsequently reported eye problems. One went on to develop severe conjunctivitis; because she alone of the three had not been wearing glasses, some investigators speculated that the lenses might have afforded some degree of protection to the others, perhaps by partially blocking ultraviolet radiation.

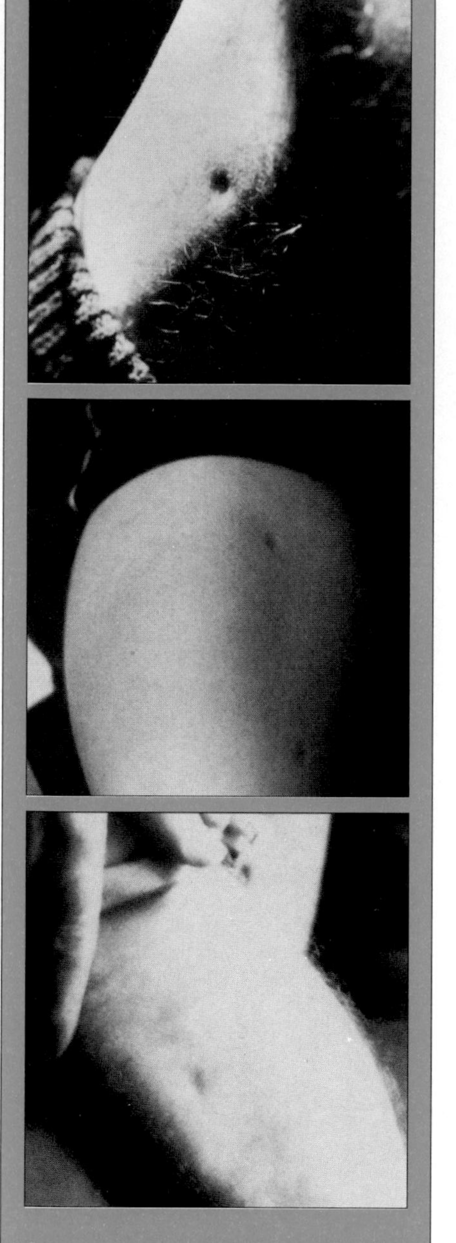

On the morning of August 17, 1988, UFO investigator Raymond Fowler found the circular, shallow depression pictured at top on his right shin. Painless yet about half an inch in diameter and a quarter-inch deep, the mysterious skin-covered indentation resembled scars exhibited by many alleged victims of physical examinations by aliens. Similar scoop marks appeared on the limbs of abductees Kathie Davis (center) and Jack Weiner (bottom) that they, too, were unable to account for. "If I had hurt myself," Weiner said, "I would have noticed."

The emotional havoc wrought by the abductions could be even more devastating than the physical effects. At the very least, abductees generally suffered from depression, reporting such effects as sleeplessness, anxiety, and occasional panic attacks. Some developed temporary amnesia. Recurrent dreams in which aspects of the abduction returned to haunt the sleeper were also frequently mentioned.

Psychologists reading through such lists quickly noted a familiar pattern. The symptoms displayed were markedly similar to those shown by combat veterans and by the victims of airplane hijackings, near-fatal accidents, and other deeply disturbing experiences. In those cases the resultant condition—marked by recurrent flashbacks to the original cause of the distress, inability to concentrate, extreme jumpiness, and a general sense of alienation and detachment—was normally diagnosed as post-traumatic stress disorder (PTSD).

A complicating factor for abduction victims was the fact that the root of their complaint was not as immediately evident as, say, rape or military combat. One psychologist writing in the *Journal of UFO Studies* attempted to address this fact by suggesting that abductees should be categorized differently from other PTSD sufferers. He proposed grouping them instead with victims of such other hidden injuries as childhood sexual abuse and psychological torture, under a separate label: experienced anomalous trauma (EAT). He pointed out that because EAT victims have a hard time convincing others of the reality of their condition, some might even develop psychosomatic symptoms as a way of validating the authenticity of their pain.

In fact, few people who investigate alleged abductees have any doubts as to the traumatic nature of the experi-

A support group of people who believe they have been abducted meets with ufologists at the New York home of artist and UFO investigator Budd Hopkins (third from left). "We look at each other and we say, 'I can't believe her story, I can't believe his story, I can't believe your story, I can't believe my story,' " said a man in one such group. "And yet, there's a comfort which we still share because behind it all, it's all the same. We don't understand it, but something happened."

ences described, regardless of whether the investigators are prepared to accept that an abduction really happened. This is particularly true of the events connected with the examination—the episode in the abduction experience that seemed to leave the most enduring memories and rouse the deepest fears.

A common element among many of the stories was the insertion of needlelike devices into the captive's body. In this respect, Betty Hill's account was to be the prototype for experiences reported by many subsequent abductees. In 1987, for example, a thirty-four-year-old Californian remembered under hypnosis an abduction that had occurred at the hands of five-foot-tall aliens three years earlier as he was preparing for bed in his parents' home in the San Fernando Valley: "I looked up and they were putting this needle into my stomach. It was really long, maybe two feet long, and it looked too fat to put into me. As it went in, I started to panic." The victim was calmed by his abductors, however, and a dark fluid was injected into his navel without his experiencing pain. Because such reports so closely resemble Betty Hill's story, some observers speculate that the accounts may reflect subconscious memories of the film version of *The Interrupted Journey.*

Other aspects of the examination have a disturbingly sexual orientation that lead critics to see in them an expression of the victims' normally repressed psychosexual anxieties. Some accounts seem to hint at deep fears of sexual invasion; victims talked of large wire objects being used to probe the rectum or of wires being inserted into the urethral tube. One abductee claimed that he turned his head from the examination table and found himself watching a replay of his past sexual encounters on something that resembled a television screen. At least two females said they were raped by their abductors.

There is, however, another interpretation of the marked sexual emphasis of such accounts. Those who assert the objective reality of the abductions see in it evidence of the aliens' fascination with the workings of the human reproductive system. In the words of Budd Hopkins, a leading proponent of this view: "The central focus of the entire UFO phenomenon is the 'study and laboratory use' of human beings with special attention to our physical, genetic and reproductive properties. . . . Apparently the center of this ongoing genetic experiment is a systematic attempt to create a hybrid species, a mix of human and alien characteristics." In Hopkins's view, men have been abducted primarily so that sperm samples could be taken from them, women to be made pregnant. The resulting embryos have been removed during subsequent abductions. Hopkins cites in support of this fantastic-sounding thesis the evidence of his own case studies, as reported in his books.

In Hopkins's 1987 work, *Intruders,* he details just such a case over a span of three generations. Its principal subject, an Indianapolis woman to whom he has given the pseudonym Kathie Davis, claimed to have experienced a phantom pregnancy that ended mysteriously. Under hypnosis she recalled not only that the baby was removed from her body by alien beings but that she underwent a subsequent abduction in which she was given a baby to hold as an apparent exercise in bonding.

As unlikely as Hopkins's theory might sound, it has been supported in the accounts of other abductees. A California artist expressed under hypnosis the conviction that her abductors were trying to extract her ova: "Oh! That's what they did. The probing was taking the egg. They're going to try and reproduce us," she cried. A gynecologist identified the surgical procedures described in some abductees' reports as laparoscopy, a process of ova removal used in the conception of test-tube babies.

etty Andreasson even claims to have been an eyewitness to the aliens' breeding program in action. In one of her many regressions under hypnosis, Andreasson recounted seeing large-headed aliens removing a small, strange-looking fetus from an apparently human woman. They then gagged its mouth and inserted needles into its cranium and ears, explaining that they could not allow it to take a breath of air. For reasons they did not elucidate, they also cut its eyelids, then placed it in a glass jar filled with liquid, the jar's cap sprouting tiny, hairlike filaments illuminated by sparks of light. The aliens referred to the process by which the baby had been born as "splicing." Subsequently Andreasson was to describe seeing eighteen-inch-long infants playing in the vivarium she visited aboard the spacecraft.

Another feature linked by Hopkins and other investigators to genetic experimentation is the implanting of small devices in the abductee's body or brain. Victims typically state that a tiny metal bead or burr was inserted into one of their nostrils with the aid of a wire probe. Betty Andreasson claims to have memories of the aliens retrieving from her body one such object, which had tiny wires implanted in it. But an attempt to have her describe the original implantation under hypnosis failed, apparently because she found the experience too painful to relive.

Adherents to the genetic-experiment theory consider such devices a means by which the abductors can keep in regular contact with their victim, who thereby becomes, in Hopkins's expressive phrase, a "tagged animal." An obvious implication of the tagging concept is that victims will be visited again by those who originally marked them out. Investigators found individuals who believe they were singled out in just that way.

One such is a businesswoman named Susan Ramstead. Her first UFO experience came in 1973. While driving to a conference, she noticed a craft that had landed in a field. Four years later, she recounted the story to an investigator, who suggested that she undergo hypnosis. Hypnotic regression subsequently revealed that she had been abducted and had been strapped naked to a table and subjected to a physical examination.

Six more years passed before the investigator heard from Ramstead again. This time she reported recurrent dreams of her earlier abductors. She saw them coming for her in her bedroom, floating her out of the house and back to the spacecraft and the examination table. Now, though, she was assailed by vivid images of metallic probes entering her nostrils.

Further hypnotic sessions revealed that Ramstead had experienced a pattern of repeated abduction in which just such incidents occurred. Ramstead's conclusion was that the aliens choose certain individuals to be human guinea pigs and then return periodically to review their progress.

An otherworldly creature holds a liquid-filled case containing a fetus that is part human, part alien in this horrific scene drawn by Betty Andreasson while recalling an abduction under hypnosis in 1987. During a previous abduction, Andreasson remembered, she had watched helplessly as aliens pinned a pregnant human mother onto a table and removed the fetus. The woman's anguished, pleading face had been appearing in Andreasson's dreams ever since.

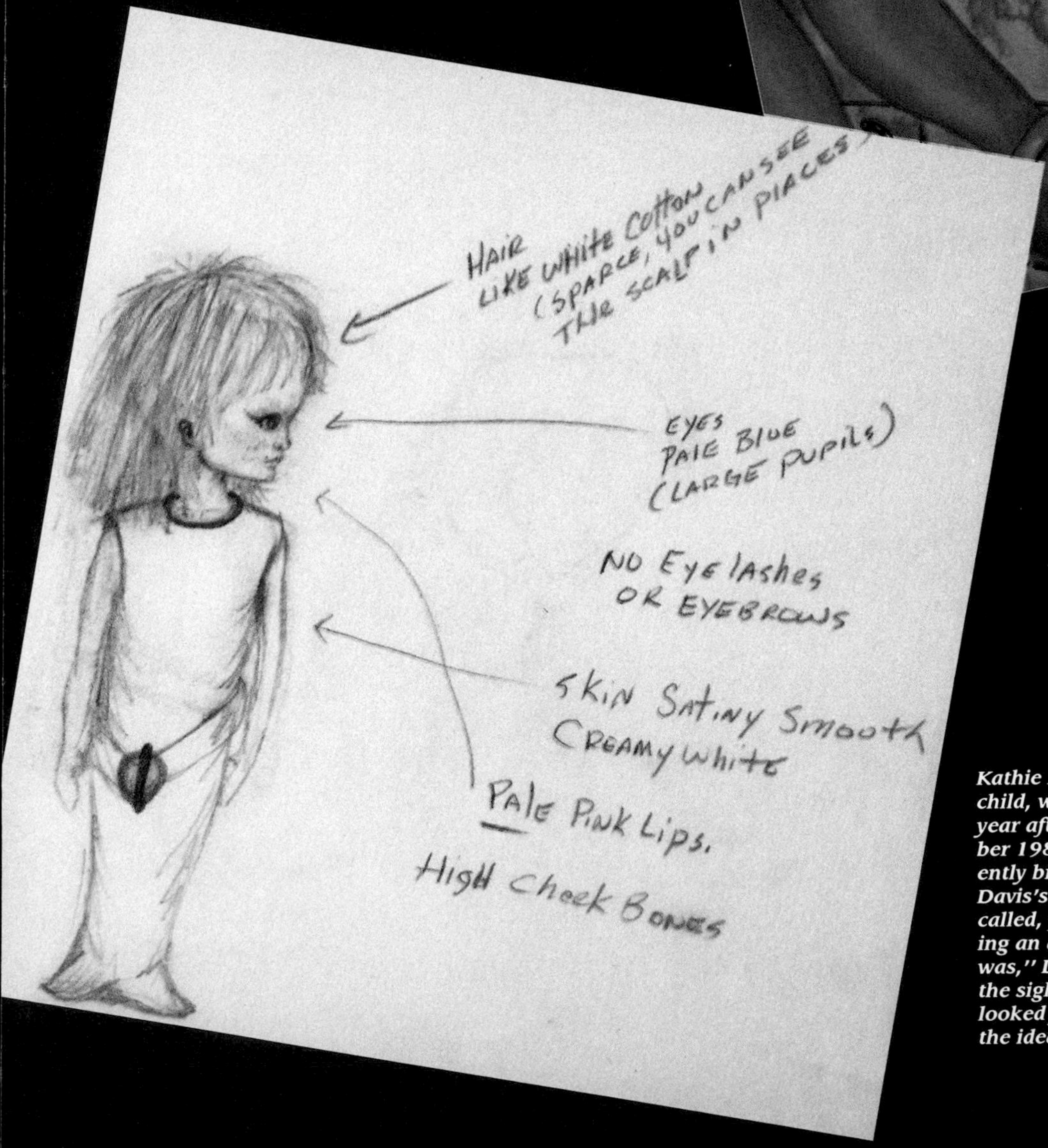

Kathie Davis said that this hybrid child, whom she drew more than a year after waking up in tears in October 1983, is her own daughter, apparently bred from an egg removed from Davis's body. Alien beings, she recalled, presented the girl to her during an abduction. "She knew who I was," Davis said sadly, "but I think the sight of me frightened her. She looked frightened, almost shocked, at the idea that she was a part of me."

The homing device in her case was a tiny metallic object placed in her nose.

The story has a happy ending, for Ramstead believes that in the course of a final abduction the implant was removed. She woke up the morning after this abduction with a massive nosebleed, which she believes resulted from the removal, and it was the nosebleed that persuaded her to contact the investigator again. She claims to feel physically better since the object's removal; headaches that had periodically troubled her no longer do so, and she has lost the sensation of having a permanently blocked nose. She does not think that the aliens will trouble her again.

Such fantastic-sounding tales inevitably raise the question of witness reliability; some skeptics regard the whole abduction phenomenon as a tissue of conscious deceit. The argument that no normal individual could make up such bizarre stories was cast into doubt by the revelation that at least one supposed abductee, an Arizona homemaker who attended one of the earliest University of Wyoming conferences, had in fact done just that.

Christy Dennis became the subject of a *National Enquirer* feature story as a result of her original claims, and in the story the respected UFO investigator Leo Sprinkle was quoted as saying, "This is one of the most remarkable abduction cases I've come across." Dennis later revealed, however, that she had invented the abduction story in order to draw attention to her concerns for the future of humanity. She said she believes that UFO abduction claims represent for some people simply "unrealized dreams and aspirations. And the only way they can get any kind of satisfaction is to fabricate some sort of story to get the focus of attention that they need."

Despite this breach of UFO enthusiasts' faith, other investigators are quick to point out that most abductees have been almost painfully sincere. There seems little reason to doubt that most believe passionately in the truth of the stories they have told. Accepting this, other observers who doubt the objective truth of their reports suggest that the abductees might reflect a particular psychological state called the fantasy-prone personality. And some skeptics even suggest that the strange stories might be the result of temporal-lobe epilepsy, whose symptoms can include complex visual or auditory hallucinations followed by a period of semiunconsciousness and sometimes amnesia.

As the abductees' claims became more pressing, the question of their veracity became one of the most urgent issues in the abduction debate. The first serious psychological study of abductees was made by psychologist Elizabeth Slater between 1981 and 1983. The survey was financed by the Fund for UFO Research, and the nine individuals surveyed were in no way a random sample; they had been selected as "credible" by Budd Hopkins and his colleagues. But Slater herself was not made aware that they had any connections with UFOs until after she had completed her testing. She was merely asked to "evaluate the similarities and differences in personality structure" of the subjects and to measure their psychological strengths and weaknesses. The subjects themselves were told not to mention any UFO connection to her.

Using a battery of standard psychological tests, she determined that the subjects as a group were above average in intelligence, were sensitive to fantasy, showed indications of narcissistic identity disturbance, and suffered some impairment in personal relationships. She also mentioned that some of the subjects were "simply flooded" with anxiety. At least six of the nine, she reported, were potentially capable of psychotic episodes; they might temporarily lose touch with reality and display confused, even bizarre behavior that was highly emotionally charged.

The puzzle is this: Did the subjects' psychological problems cause them to imagine that they had been abducted by aliens? Or did their abduction experiences bring about their psychological problems? Slater seemed convinced that the first alternative was not a viable one. Informed of the UFO connection after submitting her report, she wrote an addendum stating, "The first and most critical question is whether our subjects' reported experiences could be accounted for strictly on the basis of psychopa-

These crayon-and-pencil drawings tell the story of John, a four-year-old New York boy who walked tearfully into his mother's bedroom one morning in 1987 and asked, "Mommy, why don't you love me anymore?" John's sudden forlornness was caused by events the night before, when, he says, his mother stood by, apparently paralyzed, while he was floated out of their apartment building and up to an alien spaceship that was hovering near the moon (left). On board the UFO he met two strange beings with large, dark eyes (below). One creature escorted him; the other sat at an instrument console and never looked his way, frightening him so much that his teeth chattered. After being given a physical examination, John, still in his pajamas, was returned to his bed unharmed, although even today he cannot remember how.

thology, i.e., mental disorder. The answer is a firm no."

In 1990 another survey used mailed questionnaires to compare abductees' responses in various personality tests with those of control groups. The abductees' responses revealed above-average incidence of childhood trauma, with female abductees reporting a particularly high figure for sexual abuse. Such findings of course boosted the argument of skeptics that alien abduction reports are nothing but the workings of unhealthy minds.

Along with witness reliability, the role of hypnotism in bringing to light many of the abduction stories has also been challenged. Regression techniques have played a particularly important role in missing-time accounts. Before the hypnosis sessions began, subjects have often been unaware of anything more specific than a momentary sense of discontinuity in their lives. Some critics claim that the technique increases witnesses' suggestibility and might enable the investigator to steer their recollections in desired directions. Even proponents of hypnosis admit that there is a danger of confabulation—the process by which hypnotized people sometimes fill in gaps in their recall by inventing additional detail.

In an attempt to answer these criticisms, two ufologists in California staged an experiment. They advertised in a local college's student newspaper for "creative, verbal types" to volunteer for "an interesting experience in hypnosis and imagination." They selected eight subjects, taking care to screen out any obvious UFO enthusiasts from the sample. Under hypnosis the volunteers were told to imagine that they had been taken on board a flying saucer and given a physical examination; then they were asked to describe the experience. The investigators were astonished—and no doubt distressed—by the fluency and detail of the imagined abduction tales. Their finding: "The imaginary subjects under hypnosis report UFO experiences which seem identical to those of 'real' witnesses." The experiment showed that detailed accounts under hypnosis are not the exclusive province of abductees, and they therefore do not establish the truth of abductees' stories.

Those convinced of the reality of the UFO abductions retort that hypnosis in fact occurred in only a minority of all abduction reports, that the circumstances of the imagined-abduction interviews were different from those of "real" cases, and that mere imagination could not explain away such phenomena as multiple-witness abductions and physical traces left on victims or their surroundings.

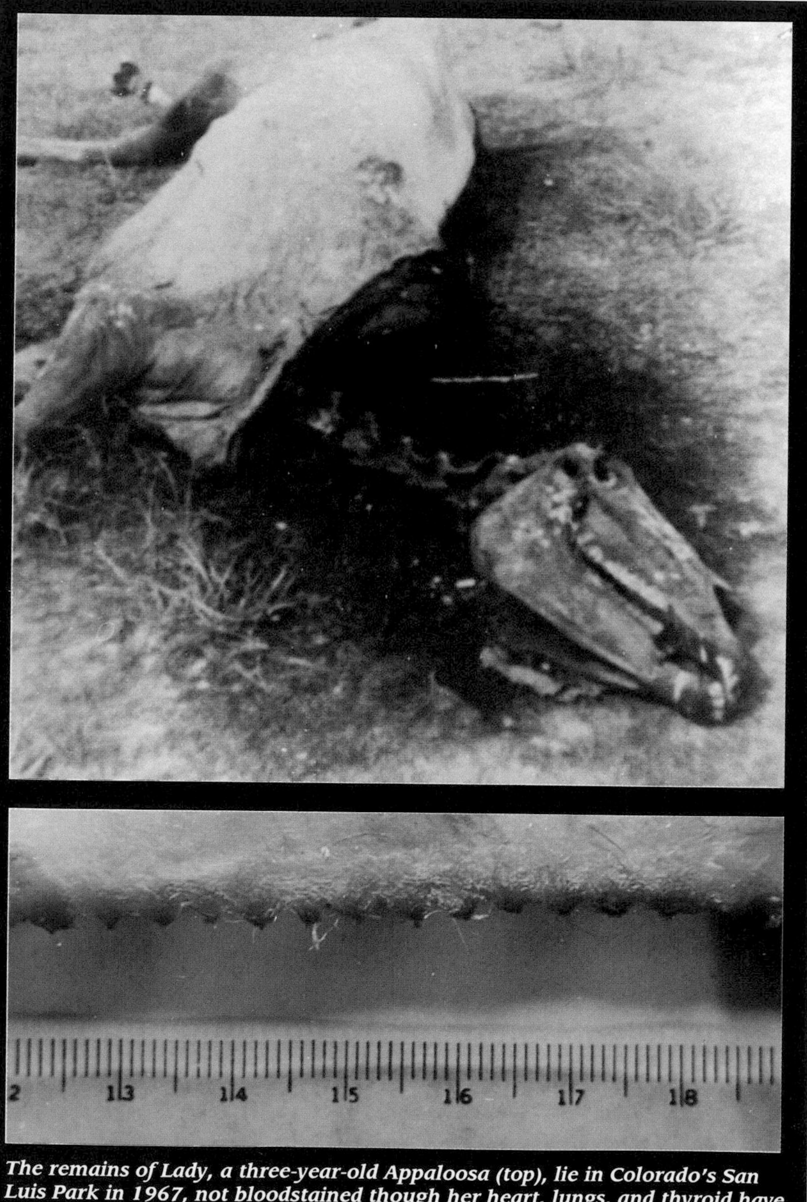

The remains of Lady, a three-year-old Appaloosa (top), lie in Colorado's San Luis Park in 1967, not bloodstained though her heart, lungs, and thyroid have been removed and the flesh stripped from her neck and head. In the lower picture, the tooth-patterned cut on the hide of a cow found mutilated in 1975 resembles the cauterized effect of a surgical laser. Some attribute to aliens a number of such livestock mutilations discovered in the western United States.

While debate raged over the abduction phenomenon, other

mysterious events that some observers linked to alien intrusions were also drawing attention. One such was a wave of animal kidnappings and mutilations that swept through the central and western United States in the 1960s and 1970s. Although the link appeared tenuous at first, voices were soon raised to claim a significant connection between abductions and animal mutilations.

Stories of cattle mutilations rose into the hundreds in 1975, by which time the problem was receiving national media attention. The spate of incidents coincided with a wave of reported UFO encounters. Putting the two phenomena together, some observers suggested that aliens may be extending their biological investigations to the animal kingdom. One Texas abductee even reported having watched the levitation of a calf and its subsequent mutilation and dismemberment aboard an alien craft.

Certainly the facts of the mutilation phenomenon at first seemed bizarre enough to encourage speculation about extraterrestrial involvement. Cow carcasses were found with body parts—often teats or eyelids—removed, apparently with surgical skill. Some were reportedly discovered in mud or snow that showed no human or animal tracks nearby. The dead animals' remains were frequently described as "bloodless," and it was implied that the natural fluids had somehow been sucked out of the bodies. Claims were made that the victims tended to decompose at unnatural rates, whether fast or slow. Oddest of all was the suggestion that predators were refusing to scavenge on the mutilated animals and that farm dogs would shy away from them.

Despite all these oddities, however, investigators examining the situation gradually arrived at earthly rather than alien explanations for the deaths. A report by the Colorado Bureau of Investigation estimated that 70 percent of the cattle had in fact died from natural causes. Veterinarians pointed out that the carcasses' apparent bloodlessness could be explained by the fact that blood oxidizes after death, losing its color—a phenomenon that could fool untrained observers. The so-called surgical removal of eyelids, teats, and other soft tissue was also disputed; the investigators asserted that these were precisely the parts that small scavenging animals would find easiest to sink their teeth into. The failure of larger scavengers to feed on the bodies could have been caused either by chance or by the diseased condition of the carcasses.

In addition to explanations found in nature, the possibility was raised of the human touch in the killings. Some investigators speculated that unnamed cultists were killing animals for use in satanic rituals. Even something so prosaic as insurance fraud might have been the motive in some of the killings: Policies on livestock would not normally cover death from natural causes, but if foul play was suspected, owners could legitimately submit claims for losses.

If a convincing alien connection to animal mutilations could not be made, a far more frightening possibility came front and center with a new line of inquiry linking extraterrestrial activity with a string of mysterious human deaths. In 1990, the French-born ufologist Jacques Vallee published *Confrontations,* a summary of his researches over the previous decade. The title was chosen to highlight the tension he now sensed in the relationship between human subjects and the extraterrestrials he believed had contacted them. For evidence he pointed to some odd events in Brazil that seemed to support his contention that UFO activity might be taking a malign and dangerous turn.

In the book, he discussed a cluster of deaths in 1981 and 1982 around the small town of Parnarama that were reportedly linked to UFO sightings. In each case the victims were deer hunters who had climbed up trees at night to await their prey. Witnesses claimed that the men had been caught in lethal rays projected from small, humming UFOs they compared to flying iceboxes. No autopsies were performed on the corpses, however, and they were buried quickly because of the tropical heat.

Similar attacks on people had been reported on the island of Colares, 450 miles to the northwest, four years earlier. That time a local doctor, Wellaide Cecim Carvalho de Oliveira, had had a chance to examine the victims. In all she saw thirty-five patients, some of whom subsequently died of their injuries. The symptoms she described included reddening of the skin, burns, and hair loss—classic indications of radiation sickness. She also reported finding small puncture marks on some patients and a decrease in the number of red blood cells.

Dr. Carvalho's report attracted enough attention to arouse the interest of the Brazilian air force. According to Vallee, a team of more than a dozen investigators was dispatched to the area; after a ninety-day period of surveillance, the team produced a 500-page report for the armed forces headquarters. No details of its contents were released, however, so once again evidence of a UFO connection could not be substantiated.

Even more tantalizing was Vallee's account of his investigation of a mysterious double death that occurred in 1966 on a Rio de Janeiro hilltop. The victims were men in their thirties, both electrical technicians from the town of Campos, 150 miles to the northeast. They were married, with families; their business had involved work with television transmission equipment.

The bodies, neatly dressed in suits and new raincoats, were found side by side. There was no obvious sign of violence. Beside them lay a strange selection of items: two crude homemade metal masks, a crushed piece of aluminized blue-and-white paper, and some cellophane soaked in a chemical substance. There were also slips of paper bearing handwritten notes. One contained elementary electrical formulas. Another read: "Meet at the designated spot at 4:30 p.m. At 6:30 p.m. ingest the capsules. After the effect is produced, protect half of the face with lead masks. Wait for the prearranged signal."

Subsequent investigation revealed that the men had apparently belonged to a spiritualist society with interest in the paranormal; one of its objectives was rumored to be communication with other planets. The police learned that the two had previously participated in experiments that involved the explosion of homemade bombs.

On the day of their death they left Campos by bus in the morning, arriving in the Rio suburb of Niterói at 2:00 p.m. There they bought the coats—it was raining at the time—and a bottle of mineral water. They started up the hill on foot at 3:15 p.m. They were last seen around 5:00 p.m. The bodies were found three days later at a secluded site about 650 feet below the summit of the hill.

The coroner's report stated that the skin of the corpses was pinkish in color and that there were possible burns, although the state of decomposition ruled out any certainty. Two separate autopsies failed to find any trace of poison. The official cause of death was filed as cardiac arrest.

UFO activity had reportedly been observed in the area a few months before the incident; a fifty-four-year-old manager and his family had told of seeing an elliptical luminous object hovering at a height of a hundred feet. More significantly, perhaps, there were reports of sightings on the evening when the men had died. One woman described watching an oval object hovering over the hill, rising and descending for three or four minutes. A line of fire rimmed its edges, she said, and a blue ray shone down from it. Other witnesses subsequently confirmed the sighting.

One of the two men was said to have spoken a few days before his death of a "final test," after which he would say whether or not he was a believer in the paranormal. There seemed to be strong reasons for believing that they expected to achieve some sort of connection, but obviously their secret died with them.

Only one conclusion can be drawn with any certainty from these mysterious matters. The more investigators focus on phenomena such as the Brazilian deaths or Hopkins's biological experiment theories, the further the benign Space Brother concept espoused in the 1950s seems to recede into the background. In the words of one investigator who shares Hopkins's views, David Jacobs of Philadelphia's Temple University, "Now we have knowledge. Now the UFO phenomenon has assumed the realistic dimensions that early and incomplete abduction material only hinted at. Now I am frightened."

A mother, awaking to find her newborn child missing, frantically overturns the pillows and blankets before discovering the baby above the bed's headboard, suspended in the arms of diminutive alien beings. A thirty-five-year-old man taken on an otherworldly journey by a beautiful female creature gazes heavenward and sees nine resplendent rings, each appearing to be a multitude of tiny, radiant stars locked in orbit around a solitary point of blinding white light. A woman stirs from a deep swoon to find puncture marks on her body, and she has no memory of how they got there.

As much as these accounts read like modern-day descriptions of abductions by aliens, they are in fact tales from days long gone by. The mother's panic is part of an old Welsh legend about baby-stealing fairies. The man on an unearthly journey—pictured above beholding a heavenly choir with his angel guide—was Dante Alighieri, fourteenth-century author and protagonist of the *Divine Comedy.* And the woman with the puncture wounds was a vampire's typical victim in nineteenth-century folklore and fiction.

Some scholars say the fact that the past abounds with stories echoed in today's UFO reports is a matter of old psychological preoccupations now being expressed in symbols that are somewhat altered but still recognizable. Many believers in UFOs, however, think the similarities, including those illustrated on these pages, cut the other way instead, furnishing evidence that close encounters with aliens have been taking place since time began.

Wondrous Visions and Visitations

Most close-encounter accounts begin the same way. While driving, relaxing at home, going for a stroll, hunting, fishing, or performing some other ordinary activity, a person notices an unearthly beam of light from an unknown source, a pulsating, multicolored sphere hanging in the sky, or some other extraordinary aerial phenomenon. Often, the UFO seems to flit effortlessly through the air; at other times, it mysteriously defies gravity and hovers noiselessly, breaks into a number of identical parts and joins together again, or performs any of a number of other astonishing maneuvers, many of which bear uncanny resemblance to those executed by supernatural figures in the Bible and in ancient legends and myths. A few of these beings are shown at right.

Are modern accounts of UFO sightings and alien visitations simply old cultural traditions dressed up in new details? Some folklorists think so. But many UFO investigators believe that even biblical descriptions such as Jacob's ladder *(far right)* or the prophet Elijah's ascent into heaven in a chariot of fire are actually reports of ancient close encounters.

Encompassed in an unearthly glow, Vishnu and his consort Lakshmi soar over a pond on the shoulders of the great eagle Garuda—a creature so powerful the beating of its wings, it is said, caused the whole Earth to wobble. Modern accounts of UFO sightings include frequent descriptions of strange, awe-inspiring light and noises.

*wo luminous fairies descend from the nighttime sky toward John
ıd Margaret, protagonists of a collection of children's stories that
ere popular in England around the turn of the century. Illumi-
ıted by a light that shone from within, the fairies first appeared
s a little star. Then, like a miniature spaceship, the pair steadily
ew closer and hung in the air just over the children's heads.*

Dreaming, the biblical Jacob sees a ladder that descends from the clouds of heaven to the rocky Earth below, much like the shafts of light that aliens, according to present-day abduction accounts, employ to convey humans to and from their spacecraft. The ladder, as described in the Book of Genesis, provided passage for gossamer-robed angels of God, who used it to climb down to Earth and then return.

Fearful Creatures Folklore o Fact

A nagloper, or night walker (above), prowls the gloom in search of sleeping women and small children. Like the aliens who today are said to abduct humans, this sorcerer of folklore was believed to induce a deep stupor in his victims and in any bystanders.

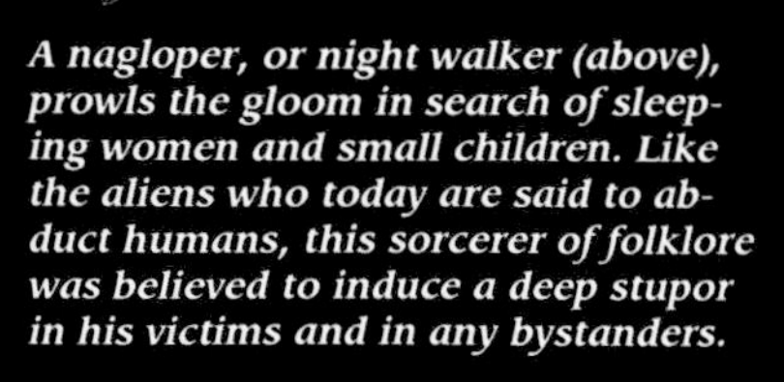

A female vampire lunges at the neck of her male prey. Like the vampire victims of legend, purported alien abductees usually can neither remember the ordeal nor account for the small wounds on their bodies that commonly result.

People who claim to have been abducted almost always report being overwhelmed by conflicting emotions. At the outset of their close encounters, they feel a short-lived sense of awe brought on by the spectacle associated with most UFO sightings. But as the realization sets in that they are experiencing not a freak of nature but something they believe is powerful, controlled by intelligent forces, and possibly threatening, childlike fascination typically gives way to unbridled fear.

Making a break for it, however, is usually out of the question. Either subsequent events unfold too quickly, study subjects say, or they discover to their terror that they can no longer control their own actions. By means of hypnosis, mental telepathy, or some other method unknown to mere humans but apparently practiced throughout the ages by the menacing creatures pictured here, the aliens seemingly rob their victims of the will to resist and draw them ineluctably away from everything they know to be familiar and comfortable.

Sent to Earth to father an evil hybrid child, a demon known as an incubus impregnates a woman while she sleeps. Some UFO investigators believe space creatures are abducting human beings today for a similar purpose—breeding offspring that are half-human, half-alien.

Mythical winged sirens sing while Odysseus and his crew ply the waters below. The sirens were irresistible to everyone who heard their song and thereby lured many a sailor to his ruin. People who say they have been in the presence of alien creatures frequently describe an identical sense of helplessness and inability to resist.

Snatch Squads and Kindly Helpers

According to any number of accounts, modern close encounters are not usually with lone alien beings but involve groups of several creatures who seem to operate with the precision of a team of interstellar commandos. The beings generally wear uniforms and all behave in basically the same manner, although in most cases, each one appears to have been assigned a specific task, which he or she goes about in a mechanistic, often brutally efficient manner. Nonetheless, most of the humans who claim to have received such harsh treatment insist their experience was not entirely bleak. Typically, this is because there is a seemingly compassionate alien among the hardhearted ones who communicates with the abductees, escorts them from compartment to compartment within the spaceship, reassures them, and provides other comforts that in earlier times were counted as blessings from a guardian angel *(right)*. Of course, not all of the people who are thought to have been abducted by aliens return to tell of finding such a protector. Like babies snatched by fairies or sinners in the clutches of demons, these poor souls seem simply to disappear, never to be seen again.

As told in a cautionary tale penned more than a century before Dante's Divine Comedy, the soul of Tondal (below), a sinful knight, stands beside his guardian angel, who has come to escort the knight on a fantastic journey through the realms of hell, purgatory, and heaven, and then back to the world of the living. Several modern abduction accounts contain descriptions of comparable treks made in the company of supernatural custodians.

ur fairies (left) spirit away a baby they found lying attended in a cradle. Required to pay a tithe to hell ce every seven years, fairies were said to sometimes 'er captive humans as payment. Like the demon below, own carrying a lost soul to damnation, the fairies can aterialize and vanish in the wink of an eye and often ss whole through physical objects—abilities they seem share with the occupants of supposed alien spacecraft.

Enchanted by the fairies' music, a hapless mortal is caught up in their whirling dance, which takes place within a ring of mushrooms similar to the circular landing traces thought to be left by UFOs.

Heaven, Hell, and the Fairy Realm

Nearly half the people who say they were abducted claim their alien captors subjected them to ruthless, often painful physical examinations. During these tests, they were cut, sexually violated, and made to suffer an assortment of other ghastly torments, some of which resembled the punishments that many believe await sinners in eternal damnation. Two such conceptions of hell are shown at right. At times their exams were followed by a goodwill tour of the spaceship and a high-speed joyride through the Earth's upper atmosphere or to another world. In rare cases, the abductees—like the ecstatic souls depicted on the following pages—were also treated to visions of a shining celestial paradise and heard a voice they thought was God's own.

Returned to the Earth at the conclusion of their encounters, most abductees have no conscious memory of the experience but are nagged by a sense of missing time. In this, they are just like humans who have been released from a fairy dance *(left)*. According to legend, most who gambol with the little people believe they have been doing so for only a matter of minutes, or a few hours at the most, when in fact seven years and sometimes more have mysteriously slipped by.

Humanoid devils thrust sharpened pikes through the bodies of persons condemned to hell for anger, one of the Seven Deadly Sins. The grisly fate depicted in this fifteenth-century woodcut closely resembles the physical examinations that many abductees say they have undergone at the hands of needle-wielding alien creatures.

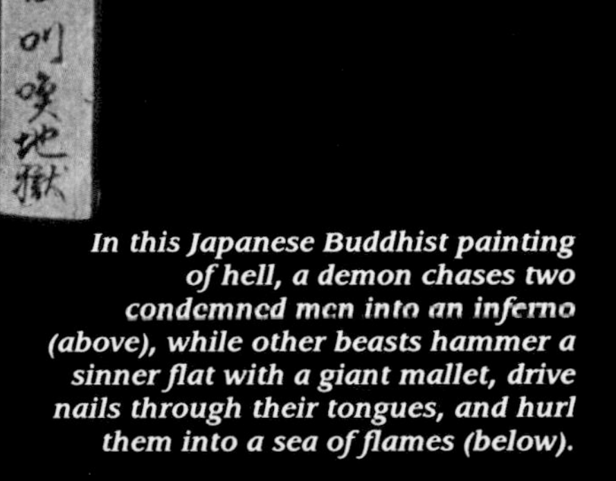

In this Japanese Buddhist painting of hell, a demon chases two condemned men into an inferno (above), while other beasts hammer a sinner flat with a giant mallet, drive nails through their tongues, and hurl them into a sea of flames (below).

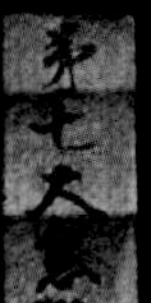

A wingless soul joins hands with angels (left), while two other spirits glide toward heaven's light-filled gate. Though painted in the 1400s, the portal is similar to the radiant tunnels described by abductees and the survivors of near-death experiences.

The Eight Immortals of ancient Chinese legend, traveling by horse, flying carpet, and other exotic means, cross the seas to paradise. Admitted to eternal life as a reward for their good works on Earth, they travel throughout the universe at will.

CHAPTER 2

Closing In on the Truth

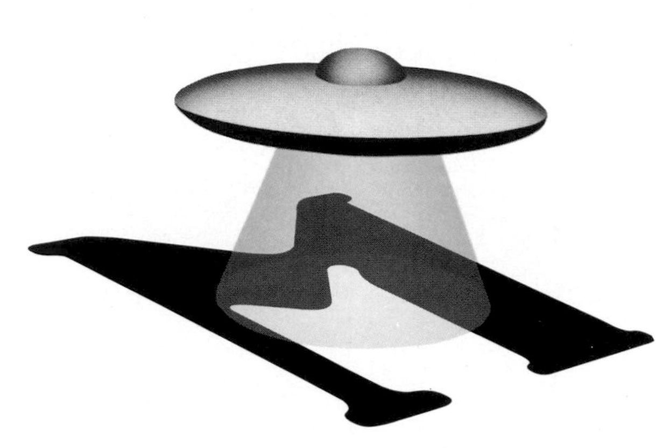

onique O'Driscoll had always thought herself a clearheaded person, not the sort to hallucinate or get involved with UFOs. At thirty-seven, she was a single parent making ends meet as an employee of the town of Brewster in suburban New York; everybody regarded her as pleasant, intelligent, and industrious. Yet what Monique O'Driscoll experienced on the night of February 26, 1983, sounded like some kind of wild, demented dream.

The event took place in midevening, related O'Driscoll. She and her fifteen-year-old daughter, also named Monique, were driving home from dinner out when the daughter cried, "Mom, look at those lights." Her mother saw them. The lights were moving slowly, just over the treetops, keeping pace with the car as she slowed to about fifteen miles per hour. The lights were very bright and multicolored—red, white, blue, yellow—perhaps fifty of them in a boomerang pattern and affixed to a dark, wing-shaped mass.

Curiosity overcoming her fear, O'Driscoll stopped the car by a pond and got out. The lights were hovering about 30 feet over the pond, reflecting off the ice. And now they began flashing in sequence up and down the winglike structure. "My daughter was screaming for me to come back," recounted O'Driscoll, "but I stood there just amazed, looking at this thing just a stone's throw away from me. It was huge." She estimated it measured about 200 feet from wing tip to wing tip. The girl was still shrieking for her mother to come back, and O'Driscoll started walking toward her car as the object slowly passed directly overhead. "I looked up and was dazzled by the lights," she said. "I shielded my eyes and saw this huge, very dark gray metallic material with all kinds of dark grill-work with pipes and tubes running up and down it. They all seemed to be connected to some type of circular-like depressions in the middle of it."

She watched the object hovering over her head for perhaps ten minutes. "The lights were flashing like crazy," she said. "Then the object made a slow, tight turn as if turning on a wheel and drifted slowly toward the east just over the trees. Then it was gone. It just vanished." There was another witness, too. A woman named Rita Rivera was parked by the pond several hundred feet away from the O'Driscoll auto. She said she saw it all, watching terrified from inside her car. And at least ten other residents in the same area reported a light-beaming, boomerang-shaped object gliding silently over their homes between 7:00 and 10:00 that same evening.

It was not the first time a UFO had been seen in the vicinity of Brewster. Two months earlier, just before midnight on New Year's Eve, Edwin Hansen, a fifty-five-year-old warehouse foreman, noticed some curious lights that first appeared to be those of a helicopter but later took on the shape of an immense and menacing boomerang. In the weeks following, a number of similar reports came in to the police—who shrugged them off as aircraft lights.

But February 26 was the first time so many people in the area had experienced the same sort of UFO at more or less the same time. It marked the start of a series of sightings that would involve far and away more witnesses than any previous phenomenon in the annals of ufology. Before the Hudson Valley sightings ran their astounding course, perhaps 5,000 people would say they observed the flights of a great, light-flashing boomerang—or boomerangs. Critics expressed doubts and offered explanations, but the phenomenon would stubbornly resist attempts at rational interpretation and thus would contribute mightily to the growing mystery of UFOs.

Indeed, in the 1970s and 1980s, as researchers worldwide refined their investigative techniques and new data piled up, an increasingly complex and disturbing appraisal of UFOs developed. The mesmerizing tales of alien abduction and experimentation would capture the blackest headlines. Yet there would be much more of compelling interest to ufologists. Investigators would cut through the plethora of hoaxes and simple misperceptions to accumulate evidence of a technology even more magical than previously believed. They would analyze numerous confrontations with UFOs, some of which would be comical were they not so eerie and in some cases ugly, even deadly.

All of this would be accompanied by more and more sightings of beings at the helms of the UFOs—until descriptions would emerge of humanoid space travelers in a remarkably diverse assortment of sizes and shapes. Researchers would study intriguing new photographs purported to be of UFOs and examine physical signs of their supposed presence on Earth. On the basis of new information, an old, once-discounted tale of a UFO crash would bear reexamination. And along with new testimony, this study would lead researchers to suspect that the United States government, for all its officially averred lack of interest, knows a great deal more than its agencies are telling—or would care to see revealed.

The citizens of Westchester and six other counties north and east of New York City might well feel entitled to any serious explanation the authorities could offer. The February 26 sightings paled in comparison with what happened about three weeks later,

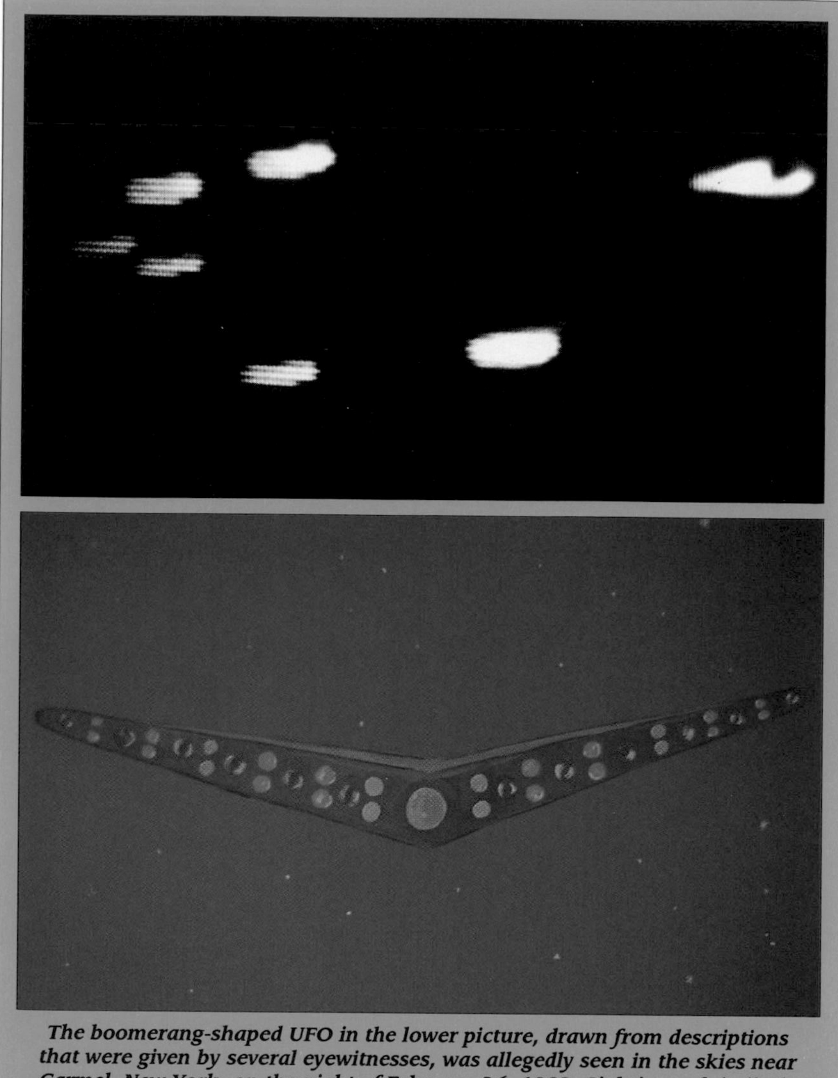

The boomerang-shaped UFO in the lower picture, drawn from descriptions that were given by several eyewitnesses, was allegedly seen in the skies near Carmel, New York, on the night of February 26, 1983. Sightings of similarly shaped unidentified flying objects were reported by more than 5,000 individuals in the Hudson River Valley area between 1982 and 1986, including the one that was supposedly captured on film in the July 24, 1984, photograph at top.

on March 17. That night, hundreds of citizens in Brewster and nearby Danbury, Connecticut, and motorists on Interstate Highway 84 frantically telephoned the police to report a large object moving slowly overhead and beaming down fantastic colored lights.

Another week passed, and then on March 24, in a two-and-a-half-hour span starting at 7:30 p.m., an estimated 2,000 people witnessed an identical phenomenon. Observers with binoculars saw a dull green structure connecting the lights and said that when it turned, they could see that it had a V shape. At one point, two sightings occurred simultaneously fifteen miles apart in Millwood and in Yorktown, where more than 1,000 people witnessed V-shaped lights flying slowly at altitudes between treetop level and 2,000 feet, occasionally stopping to hover over autos and individuals, as if inspecting them. Some people said the UFO was the size of an aircraft carrier; to others, it looked like "a flying city." A few witnesses associated a faint whooshing sound with the UFO, but most thought that it was silent. "The strange thing about it was that the object made no sound," said Ruth Holtsman, who was in a car with her family at the time. "It just hung there motionless in the sky. It was like seeing a ghost."

Then, the uproar died down. More than a year passed with only sporadic sightings. But on June 14 and July 24, 1984, the boomerang allegedly reappeared, this time seeming to reconnoiter the Indian Point nuclear power plant near Buchanan, New York. On the fourteenth, a dozen guards at the plant said they saw what one of them called "ten or more lights arranged in a boomerang pattern" hovering a quarter of a mile away for about fifteen minutes. On the night of the twenty-fourth, according to guards, an object again appeared over Indian Point, its lights in a semicircle first flashing yellow, then white, then blue, with a blinking red light to the rear. As the object approached to within 500 feet of the guards, they could make out a shape like an ice-cream cone. It seemed gigantic, as big as three football fields and going so slowly that a walking man could keep up with it.

Unofficial reports later said that as the UFO came nearer to one of the reactors, an officer inside the plant started filming with a remote video security camera atop a ninety-five-foot pole. The object was so immense that the officer had to pan the camera 180 degrees in order to catch it all. It was said that during the UFO's approach, the plant's movement-detecting sensors failed, as did the rest of the alarm system and the computer programmed for security and communications. Finally, after about twenty minutes, the UFO passed over some trees and out of sight.

According to local rumors, authorities clamped a lid

on the episode the next day. The guards were ordered to forget what they had seen, the video records were removed, and officials of the U.S. Nuclear Regulatory Commission shortly oversaw a shake-up of the plant security operation. Sightings in the suburban New York area continued through August, with six that month, then began to diminish. The characteristic boomerang pattern of lights was sighted again in October, twice more in 1985, and three times in 1986. By 1987, the phenomenon seemed to be at an end.

If for no other reason, the Hudson Valley sightings would be important for the sheer number of witnesses and the consistency of their reports. The principal investigator was Philip J. Imbrogno, a well-known area ufologist who led a team working in conjunction with J. Allen Hynek's Center for UFO Studies. Imbrogno started his research in 1983 at the height of the tumult. He and his colleagues interviewed no fewer than 300 witnesses to the March 24, 1983, episode alone.

The investigators sought out people they thought most likely to be competent observers—and least likely to succumb to mass hysteria: doctors, lawyers, engineers, scientists, corporate executives. An engineer trained in air-traffic control systems watched five green lights in the shape of a boomerang moving silently through the sky. The National Weather Service's chief meteorologist at Westchester County Airport said he saw the V-shaped object moving slowly overhead, its lights changing color "as if there was a rotating prism in them." He estimated its size at 1,000 feet from wing tip to wing tip. "This was not any type of aircraft I was familiar with," he told Imbrogno. Nor was it, he said, "any type of formation of aircraft."

However, a formation of aircraft was precisely what a number of UFO debunkers made of the Hudson Valley sightings. In November 1984, in the midst of the excitement, the popular science magazine *Discover* published an article claiming that it was all a gigantic spoof perpetrated by a group of fliers in small aircraft from the airport at Stormville, New York. The Stormville Flyers, as the magazine admiringly dubbed them, had equipped their planes with colored lights and had gone aloft at night to dazzle and befuddle the citizenry with their flashing beacons and formation flying.

oreover, there was some corroborating evidence to ponder. A pair of UFO investigators, Dick Ruhl and Richie Petracca, working for the respected Aerial Phenomena Research Organization (APRO), announced that they themselves had just witnessed a performance by the Stormville pilots. As Ruhl recounted the incident, he and Petracca were driving on Interstate 84 at 9:30 one evening when they spotted "a brilliant white wedge-shaped object floating and turning in the sky." Then another object came into their view, and the two UFOs glided slowly overhead, maneuvering around each other and flashing multicolored lights. Apparently, further objects appeared at some point, for Ruhl related that "they finally formed up in a huge boomerang shape and it was then that I saw some light reflected on the bodies of six aircraft. We knew we had evidence of the 'Stormville pilots.' "

Yet that did not satisfy investigator Imbrogno and his colleagues that their job was done. Witness after witness had said the lights were all shining from a single dark, nonreflective metallic structure in a boomerang configuration. None of these observers, some of them aviation professionals, ever reported instead seeing individual aircraft flying in

IBM program manager Ed Burns demonstrates to investigators Philip J. Imbrogno (left) and J. Allen Hynek the maneuvers of the UFO he said he observed while driving along the Taconic Parkway in Westchester County on March 24, 1983.

This issue of Discover argued that the Hudson Valley UFOs were actually small planes flown in formation. A drawing (far right) showed how such a formation of aircraft with their exterior lights on might mimic a UFO. By dousing those lights on cue and holding position with the help of their cockpit lights (near right), the pilots could make it seem the UFO had suddenly vanished.

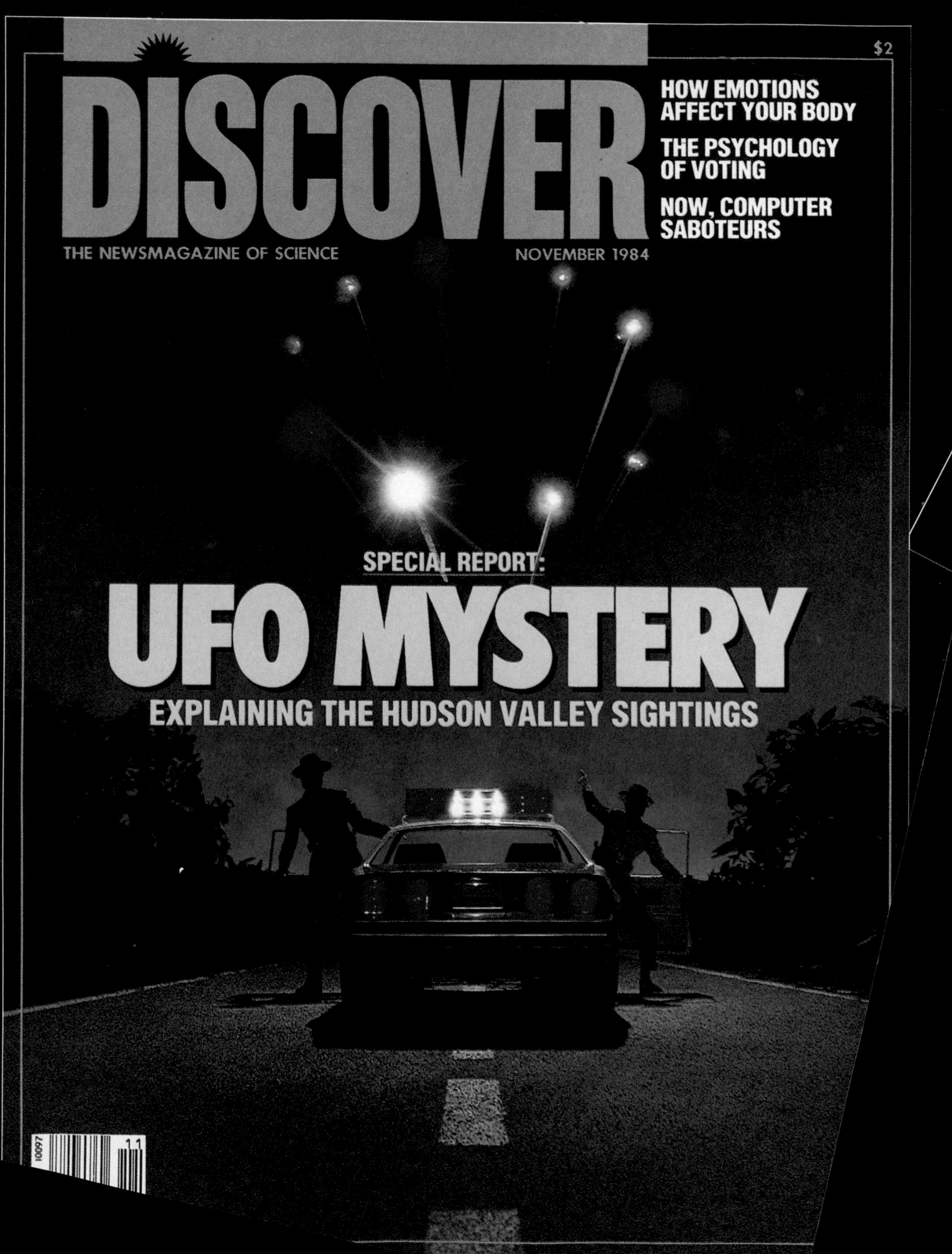
$2
DISCOVER
HOW EMOTIONS AFFECT YOUR BODY
THE PSYCHOLOGY OF VOTING
NOW, COMPUTER SABOTEURS
THE NEWSMAGAZINE OF SCIENCE
NOVEMBER 1984
SPECIAL REPORT:
UFO MYSTERY
EXPLAINING THE HUDSON VALLEY SIGHTINGS

formation. Imbrogno, Hynek, and their colleagues conceded that some Stormville pilots may well have had some fun imitating UFOs. But in Imbrogno's view, that could not and did not account for more than a minuscule fraction of the Hudson Valley sightings. Wrote Imbrogno in a paper presented to a Mutual UFO Network (MUFON) symposium in San Antonio, Texas, in July 1984: "It is my belief that we are dealing with a large object of unknown origin. It seems also that at least two or three other objects were out that night"—March 24—"causing panic around the area." He added, "Most of the witnesses feel insulted by the aircraft theory."

Whatever the Hudson Valley sightings represented, the phenomenon was benign or at least not overtly threatening. Yet UFO records have increasingly included a confrontational element, often with an air of capricious deviltry—even outright malevolence—about it. The confrontations have commonly included extraordinarily intense and penetrating light, a light almost liquidlike in its density. A stunning sort of electromagnetic lift effect has often reportedly accompanied the light, carrying aloft heavy objects in defiance of the laws of gravity. Machines and equipment have ceased to function; UFO case studies contain hundreds of accounts of auto engines stalling, lights dimming, radios conking out. Unlike the silent Westchester boomerangs, many UFO confrontations were marked by unbearable sonic emissions and such physical symptoms as uncontrollable shaking, difficulty in breathing, chest pains, and in some cases, paralysis and amnesia.

On January 24, 1974, near Aische-en-Refail, Belgium, a woman was driving home in her Volkswagen on a clear, crisp afternoon, when—she later reported—she spied a metallic object to the left of the road ahead. It was red in color with a flattened dome on top and two rows of ports around the circumference. As she approached, she said, her radio cut out, the engine sputtered and died, and the car coasted to a stop within ten yards of the object. After a time, the object lifted off the ground, and the car restarted itself. The witness suffered nightmares and an abnormal lassitude for several days but otherwise was unharmed.

Three weeks later, on the other side of the Atlantic, two brothers were driving a U-Haul truck from Idaho to California. According to their account, they were passing through Nevada at 4:15 a.m. when the driver was dumbstruck to see a large, round, orange object flying alongside the truck on the left as three smaller luminous objects kept pace on the right. Suddenly the orange object crossed in front of the truck, and it felt, reported one of the brothers, "like we had been hit by a blast of wind or a force field." Now, the truck headlights started flickering on and off, and the engine began to miss. The brother who was driving lost control of the vehicle, and they both saw that it was floating above the pavement and drifting off the road.

Looking out, the brothers gaped at what they de-

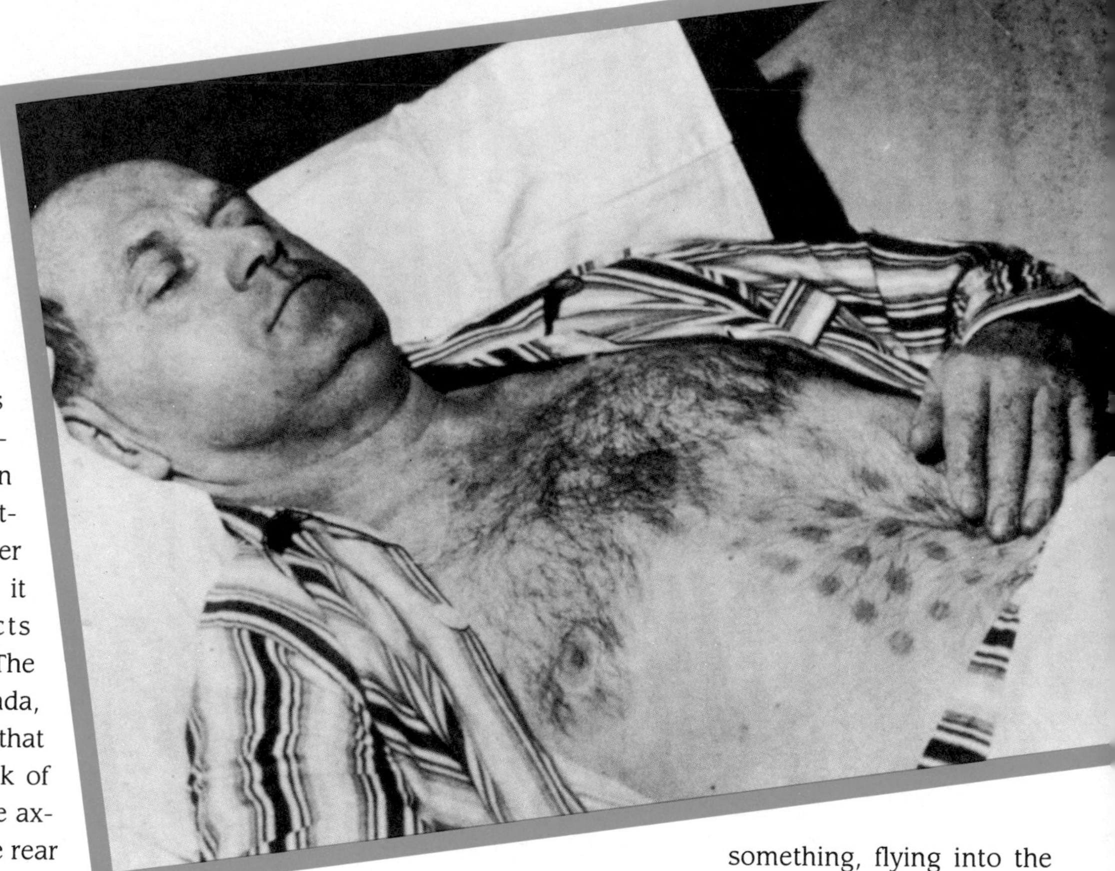

scribed as "a huge metal ball with sharp wings." The thing was coming closer and the brothers watched it with growing apprehension. "We felt that we were in a vacuum of some kind and isolated from the rest of the world." After about twenty minutes, whatever it was departed and the effects ceased. But the truck was junk. The U-Haul station owner in Ely, Nevada, who went to the scene, reported that when they tried to raise the back of the truck, they discovered that the axle had been twisted away, and the rear wheels just fell off.

All through the 1970s and 1980s similar reports kept coming in—from Denmark, Australia, Ohio, Mexico City; from a pair of Louisiana fur poachers whose straining power boat was held stationary by what they called "strong gravity forces"; and from a trio of Indiana truckers traveling in caravan whose vehicles were engulfed in a dazzling blue light that made their CB radios fail as the trucks slowed to five to ten miles an hour, the engines coughing and jerking.

Some astounding testimony came from two Chilean auto rallyists, who on September 23, 1978, were on a long, rugged road course sponsored by the Automobile Club of Argentina. Carlos Acevedo, thirty-eight, and Miguel Angel Moya, twenty-eight, were about twenty miles north of the Rio Negro in eastern Argentina and traveling sixty miles per hour when an intense light blossomed in the rearview mirror of their Citröen, growing in size so fast that they thought it must be an overtaking car. Acevedo slowed and pulled right. But suddenly, their vehicle was filled with light—"yellowish with a violet tint"—and with suffocating heat. Then Acevedo realized he had lost control of the Citröen. "I looked through the window and saw that we were about six feet off the ground," said Acevedo. "I thought we had hit something, flying into the air, and I started getting ready for the moment when the car would hit the road again."

But the vehicle continued climbing through the yellow light. So thick was the haze that Acevedo could not discern his instrument panel or Moya beside him. The two rallyists lost all sense of time until they felt a bump and realized that the car was back on the ground again. The intensity of the light began to diminish, and they saw something like a yellow cone, then an oval disappearing into the distance. "I was stunned," Moya related. "My hands were shaking and I felt something pressing against my chest. It was difficult to breathe." Nevertheless, the rallyists managed to drive on. A gas station attendant at the next town told them others had seen a bright yellowish light headed west that morning.

cevedo's and Moya's shortness of breath and shaking hands lasted for about twenty-four hours. According to ufologists, such physiological effects are common after encounters with UFOs. Nausea frequently occurs as well, and women have suddenly experienced the onset of their menstrual periods, sometimes weeks early. The effects can be much worse, however, as evidenced by what is said to have happened in Salto, Uruguay, on February 18, 1977.

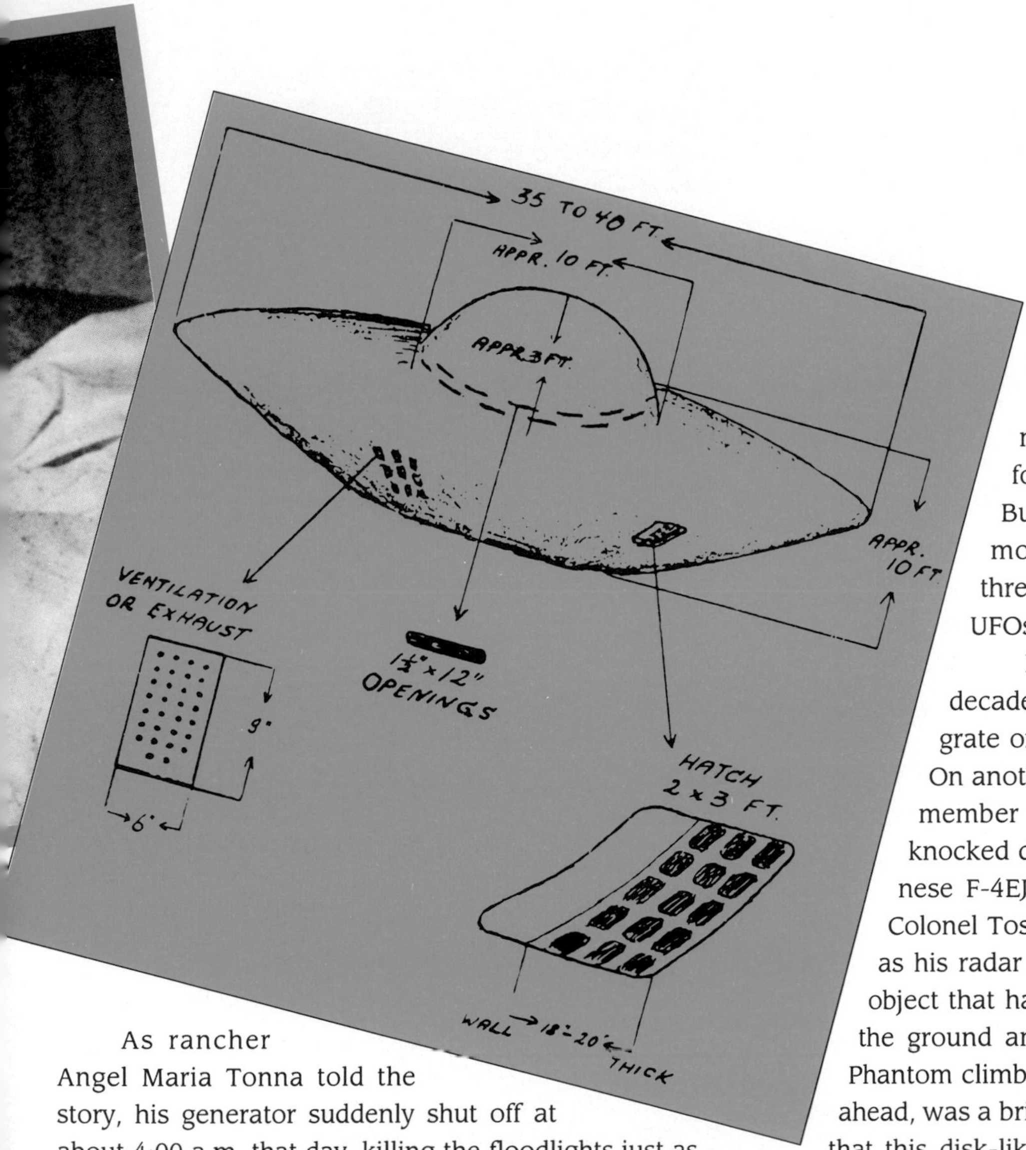

Canadian Stephen Michalak lies in a hospital after his purported encounter with a spacecraft like the one in this sketch from a Canadian UFO magazine. He said he was burned by a blast of searing air from a grille-covered opening on the UFO. The May 1967 incident reportedly also left Michalak suffering from headaches, nausea, and other disorders.

As rancher Angel Maria Tonna told the story, his generator suddenly shut off at about 4:00 a.m. that day, killing the floodlights just as the farm was starting to stir. Going out to check on the problem he saw a large disk "like two plates facing each other" hovering about twenty feet off the ground. Tonna said it was bright orange in color and illuminated the farmyard with its brilliant glow. As his sixty-pound German shepherd, Topo, ran forward barking, Tonna saw the UFO shoot out three lightninglike rays from each side. He felt intense heat and what he described as electric shocks that left him temporarily paralyzed. Then the UFO moved off, and the generator started up again yet produced no electricity because the wires had been burned out.

Tonna came out of it with severe skin irritations. But the dog refused to eat and died three days later. After an autopsy, the local veterinarian reported that the animal had suffered extreme internal heating, ruptured blood vessels, and damaged organs—as if it had been cooked to death. Assuming the veracity of Tonna and the vet, it seems too much to believe that the UFO was defending itself against the threat of a noisy canine. It would seem more likely that man and dog penetrated some sort of force field that proved fatal to the animal. But when human agencies have in fact mounted what could be construed as a threat, the reactions and powers ascribed to UFOs have been swift and formidable.

In Cuba in 1967 and in South Africa a decade later, jet fighters were said to disintegrate or vanish while apparently chasing UFOs. On another disastrous occasion a surviving crew member stated flatly that his aircraft had been knocked down by a UFO. On June 9, 1974, a Japanese F-4EJ Phantom fighter piloted by Lieutenant Colonel Toshio Nakamura, with Major Shiro Kubota as his radar officer, thundered aloft to check out an object that had been seen by dozens of observers on the ground and picked up by air defense radar. The Phantom climbed to 30,000 feet, and there, a few miles ahead, was a brilliantly shining object. "Even at first I felt that this disk-like, red-orange object was a flying craft, made and flown by intelligent beings," said Kubota later. The UFO seemed about thirty feet in diameter with square marks around its circumference that Kubota thought "might have been windows or propulsion outlets."

As the Phantom closed on the UFO, remembered Kubota, the object "dipped into a shallow turn, as if sensing our presence." Pilot Nakamura armed his 20-mm cannon and maneuvered to place the object in his gun sight. "Suddenly, the object reversed direction and shot straight at us," said Kubota. Nakamura broke hard left and rolled the Phantom into a gut-wrenching dive. "The glowing red UFO shot past—missing us by inches," recalled Kubota. "Then, it made a sharp turn and came at us again." The UFO threatened the Phantom with high-speed passes again and again—until, said Kubota, it finally crashed headlong into the jet, which exploded in flames.

Somehow, both Phantom crewmen managed to eject. But Lieutenant Colonel Nakamura's parachute pack was on fire and he perished, while Kubota floated down to safety.

Meantime, the UFO evaporated from sight—and from air defense radar screens.

Some have suggested that the two officers were so anxious to engage a UFO that they "saw" what they wanted to see, and that pilot Nakamura then overstressed his aircraft, causing its disintegration. Japanese authorities conducted a lengthy investigation, the results of which have never been made public, except for the terse statement that an F-4EJ Phantom fighter-bomber had crashed, killing Lieutenant Colonel Toshio Nakamura, following a midair collision with "an aircraft or object unknown."

Whether or not angry alien space pilots brought down the Cuban, South African, and Japanese jets, investigators generally agree that a high order of intelligence seems to govern the actions of UFOs. The search for answers to that intelligence is particularly pressing because of the growing number of abduction and confrontation stories. Moreover, there are suggestions that alien beings are willing to let themselves be seen in all sorts of circumstances.

Indeed, the MUFON study group has logged some 2,000 humanoid sightings in the last four decades or so. Many are unchecked reports. And some are regarded as pure hoaxes. One such was an October 1989 news item describing a glowing red spacecraft that landed in a park in Voronezh, 300 miles southeast of Moscow. According to the Soviet news agency Tass, which quoted three boys who observed the phenomenon, three animate objects disembarked wearing silvery coveralls and bronze boots. Two of them were creatures between nine and thirteen feet tall; they had no real heads, just bumps on top in which were set two eyes and what appeared to be a lamp. The third seemed to be a robot; while it looked like the first two, it had buttons on its chest that the others pushed to make it perform. Tass wrote that the three briefly promenaded around the park, temporarily made one of the boys disappear by pointing a long, tubular "gun" at him, then climbed back into their machine and zipped off.

The Western press hooted at the account as an absurd tabloid manifestation of the new *glasnost* freedom sweeping the Soviet Union. But one researcher argued for a different interpretation. Citing sources in the Eastern bloc, ufologist Bill Knell explained that a genuine UFO had in fact appeared that day over a highly sensitive Soviet air base near Voronezh and had hovered there paralyzing radar and other defenses until it chose to depart. The KGB did its best to silence all witnesses but reasoned that some sort of UFO story was bound to get out and therefore concocted the park fiction to mask the real event.

There are numerous humanoid reports that ufologists are inclined to take more seriously. Quite often, the beings seem to desire only communication—and by some accounts they have at times succeeded. A farmer named Gary Wilcox, near Binghamton, New York, related that he had actually carried on a dialogue with aliens. He had been out tending his cows, said Wilcox, when he saw a UFO land. He walked up and tapped it to see if it was solid—at which, two little creatures emerged and engaged him in conversation, their voices coming to him by a sort of telepathy. Among other things, the aliens predicted events in the human world, including the death of an American astronaut. Shortly thereafter, the crew of *Apollo 1,* astronauts Gus Grissom, Edward White, and Roger Chaffee, perished in a tragic fire on board their space capsule. The Wilcox encounter was subjected to a careful examination, including a psychiatric evaluation. The researchers concluded that Gary Wilcox

was sane, not a hoaxer, and that he had experienced something very strange and bewildering.

In many such meetings, the beings are said to have been entirely peaceful, even friendly. But numerous other witnesses testify to nightmarish confrontations with extraterrestrials. Californian Donald Schrum, for one, will never quite get over what he remembers of September 4, 1964. As he told the story, he was bowhunting in Placer County with friends when they became separated, and at sunset he decided to spend the night safely in a tree. After some hours, he spotted a white light zigzagging at low altitude. Thinking that it was a helicopter, Schrum climbed down from his uncomfortable perch and lit a fire to attract attention. The light turned toward him and stopped about sixty yards away. It was not like any helicopter the twenty-eight-year-old Schrum had ever seen and in fact frightened him to the point where he scrambled back up the tree.

hortly thereafter, two humanoids and what looked like a robot approached and started shaking the tree, as if trying to dislodge the hunter. At one point, the robot belched out a white vapor that made Schrum nauseous. Nevertheless, he managed to fit an arrow to his bow and launch it at the robot; there was an arclike flash and the robot recoiled. Schrum shot two more arrows at the robot, scattering his attackers. Then, a second robot arrived and more vapor rendered Schrum momentarily unconscious. When he came to, he saw the pair of humanoids trying to climb the tree after him. Somehow, Schrum managed to ward them off. The assaults continued all night. Near dawn, more beings approached, and what Schrum remembers as "large volumes of smoke" knocked him out again. When he regained consciousness, he was hanging from the tree by his belt and the UFO and its occupants were gone. A companion who also had gotten lost said that he, too, had seen the UFO that night but had not been molested.

An even stranger report came from a small lumber town named Happy Camp, also in Northern California. The encounter began on October 25, 1975, when a pair of lumbermill electricians in a truck noticed two extremely bright objects approaching in the sky. According to the two men, Stan Gayer and Steve Harris, one of the objects surged over a ridge, changing color to red and coming down toward the truck "like the lit end of a cigar," as one of them put it. The men raced away and later saw a glowing red object resting on the side of a nearby mountain.

Returning two days later, Gayer and Harris searched the area and found nothing of interest—until they peered into some bushes. There was a set of bright silvery eyes glaring out at them, and suddenly the air was filled with the sound of a siren. Prudently, the men clambered into their pickup truck and headed swiftly back to Happy Camp.

The lure of the unknown led Gayer and Harris back into the hills twice more. They were accompanied by Helen White, a sixty-two-year-old grandmother who had learned of their encounters and was as curious as they were. Her inquisitiveness was amply rewarded. On her first trip, Harris started firing a rifle at random into the bushes where the eyes had appeared. The idea may have been to flush out hoaxers, but instead the witnesses saw two silhouettes rise from the underbrush nearby. According to the story, the creatures were wearing helmets like those of arc welders and were bathed in a strange, misty light. Again, the wailing siren sounded. As the luminescent silhouettes approached, the members of the group felt a warmth and heaviness in

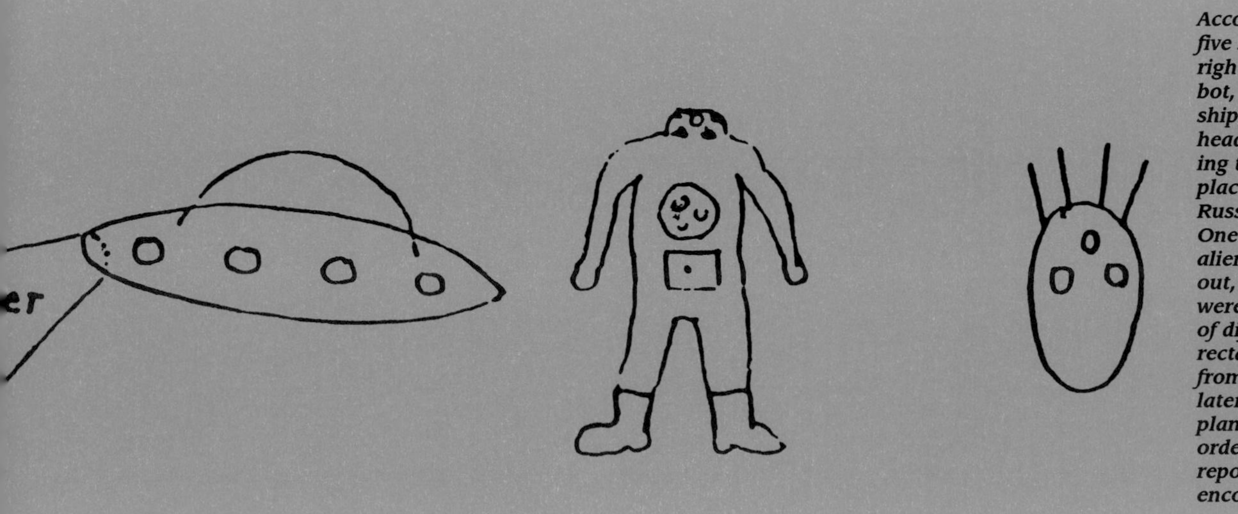

According to young witnesses, five sketches—from left to right, a spherical craft, a robot, a saucer-shaped spaceship, an alien, and the alien's head—document a UFO landing that supposedly took place in a park in Voronezh, Russia, in September 1989. One youth said that two tall aliens and the robot came out, and on the aliens' torsos were "a disk with three spots of different colors, and . . . a rectangle . . . sticking out from the body." Investigators later claimed authorities had planted the fanciful story in order to devalue whispered reports of a genuine UFO encounter that same day.

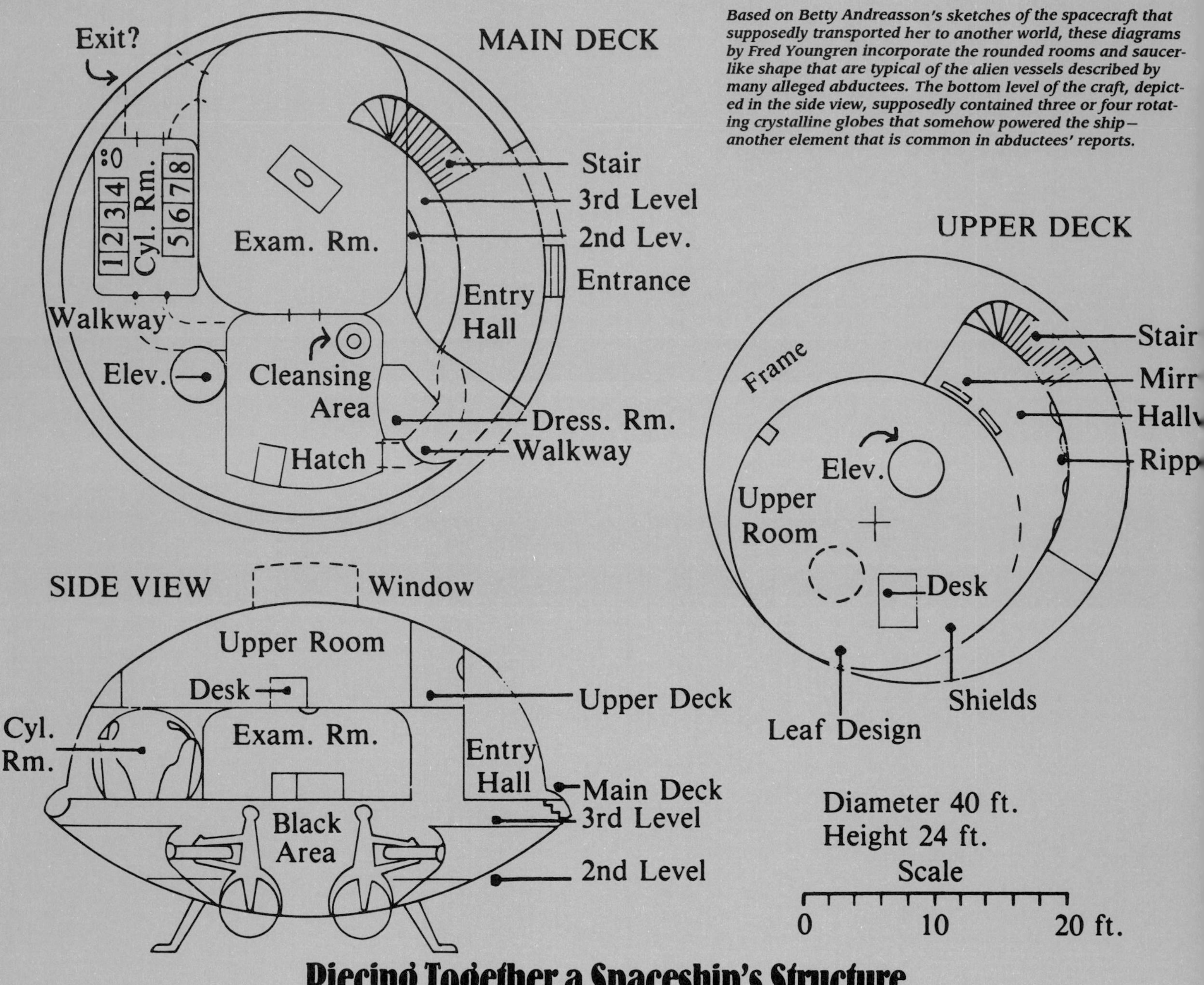

Based on Betty Andreasson's sketches of the spacecraft that supposedly transported her to another world, these diagrams by Fred Youngren incorporate the rounded rooms and saucer-like shape that are typical of the alien vessels described by many alleged abductees. The bottom level of the craft, depicted in the side view, supposedly contained three or four rotating crystalline globes that somehow powered the ship—another element that is common in abductees' reports.

Piecing Together a Spaceship's Structure

When UFO investigators interview witnesses to alleged alien encounters, they usually press for details about the visitors' spacecraft. Of particular interest are the accounts of people who claim to have been carried off to such vessels, since they can be assumed to have taken a close look at the craft both inside and out. But according to folklorist Thomas E. Bullard's statistical analysis of more than 300 abduction tales, the typical abductee passes so much of the experience inside the craft "that what he brings home is largely a Jonah's eye view, with the most explicit and most coherent descriptions applying to the belly of the whale." Recollections of the exterior of the vessels are generally limited to shape and size, he says, with most witnesses reporting disk-shaped craft.

The majority describe spaceship interiors as being divided into rooms, Bullard found, with the chamber where the victim was subjected to examination referred to most often and in greatest detail. Also mentioned have been so-called conference rooms, where the aliens supposedly interrogated captives; engine rooms; control rooms; and corridors. The rooms were usually said to be rounded, with smooth surfaces. According to Bullard, the interiors—such as that described by Betty Andreasson from her 1967 abduction experience—often held more than the ship's outside dimensions would seem to permit. Through efficient arrangement, however, ufologist Fred Youngren was able to incorporate the rooms referred to by Andreasson into his re-creation of the craft's design *(above).*

Members of the Houston-based Vehicle Internal Systems Investigative Team (VISIT) hope to use data like that collected by Bullard to one day construct a technological model of a UFO. They are particularly interested in witnesses' impressions of the lighting, sounds, air flow, temperatures, textures, and control configurations of the alien craft for the potential clues they offer about UFO engineering systems.

the air "like a sauna or steam bath, only much worse," said Harris. White said it felt like something was putting pressure on her chest. They panicked and raced down the mountain followed by a glowing red object.

Five days later, on November 2, the two electricians and White, along with two other people, returned yet again to the site. What happened is confused in their minds. But they said that when they drove into heavy fog, they apparently tried to turn back and ran into an avalanche of boulders cascading down a cliff and bouncing around the truck. "Everything seemed to happen in slow motion," said White. They remembered the truck doors, which they had been careful to lock, being unlocked as if by an unseen hand and then opened. Steve Harris said he started to reach for his gun but was somehow compelled to stop when he heard a voice tell him, "You won't need that."

Helen White recalled being lifted inside a room and having a dialogue with one of the aliens, who held up an object and described it as being gold. White replied that gold was not transparent. "There is such a thing as gold that you can look through," responded the being. "It's in your Bible." (The biblical reference, whether it was actually new to White or buried somewhere deep in her memory, was apparently to the twenty-first chapter of Revelation. There, the new heavenly Jerusalem revealed to the author was "of pure gold, like transparent glass.") The next that any of the five people remember is driving back down the mountain loudly singing a church hymn.

UFO researchers visiting the area interviewed people who said that there had been numerous previous sightings, including several by law enforcement officers who later retracted their reports. Helen White confirmed and elaborated on her experience; she said that at one point she warned an alien about the falling boulders only to have the being reply: "Don't worry, they won't hurt me." And the investigators learned that subsequent to the fall 1975 events, other people in Happy Camp had heard sirens of an ear-splitting intensity, had seen luminous fogs with humanoid figures looming weirdly on the periphery, and had observed variously shaped objects flying over the area. In one episode, two people sat slack-jawed in awe as a huge Douglas fir snapped in half for no apparent reason, while a powerful force dragged their tire-spinning pickup truck backward fifty feet around a bend. This occurred, they said, as a brilliant white sphere passed overhead.

On her first trip into the hills at Happy Camp, Helen White had the foresight to bring along a camera. But she said she was unable to photograph the glowing alien creatures, perhaps because she was choking and suffering great physical pressure on her chest.

Had she managed to use her camera, numerous questions about the fall 1975 events at that logging town might have been resolved. But she did not. It is an intriguing aspect of ufology—and one that induces strong doubts about the veracity of reports—that with all the thousands of sightings, scarcely any acceptable photos exist.

Even when pictures are produced, the images under analysis usually turn out to be birds, planes, balloons, odd cloud formations, or optical phenomena. Quirks of photography can result in wonderful-looking UFOs. Particles of dust or raindrops in front of the lens can leave convincing spheres of brilliant light on film; ordinary lens reflections occasionally will produce flat disks, and faulty development techniques sometimes result in UFO-like flaws on the finished photo. Then there are the endlessly inventive hoaxers, with their photomontages, double exposures, magnifications, and models strung on wires. Even a hubcap tossed into the air can look like a UFO on film. Yet every once in a while a picture or a series of pictures seems to defy contradiction and to offer the sort of evidence ufologists hunger for.

One of the most celebrated episodes involving photographic proof occurred in the little town of Gulf Breeze on the Florida panhandle near Pensacola. The events focused most sharply on one man, who during six months in 1987 and 1988 took literally dozens of photos, including video motion pictures, and reported some of the most bizarre en-

Taken by an amateur photographer vacationing on Canada's Vancouver Island in the fall of 1981, this is one of the few photographs of a true UFO, according to Richard F. Haines. A retired NASA research scientist, Haines analyzed the original negative using a variety of sophisticated techniques, including computer enhancement. He concluded that the image of the saucerlike object to the right of the mountaintop could not have been produced by a double exposure or by suspending or tossing an object in front of the camera. Moreover, the photographer claims not to have seen the disk when she snapped the picture—further confirmation, says Haines, that it is not a hoax.

counters with aliens in the chronicles of ufology.

At forty-one, Edward Walters was as solid a citizen as dwelt in Gulf Breeze. He was the president of a prosperous construction firm, active in the community, securely married, and the father of a teenage son and a twelve-year-old daughter. But at 5:00 p.m. on November 11, 1987, his life took on an entirely new complexion. The following is his account of what happened, mostly as related in *The Gulf Breeze Sightings,* the 1990 book he and his wife, Frances, wrote about the incidents.

It began, Walters said, while he was working in his home office, and he noticed something bright hovering just beyond a pine tree in his front yard. Stepping outside, he was amazed to see a craft in the shape of a huge child's top with portholes across the middle and a glowing ring on the bottom. "It seemed to be nearly as big as the houses below it and three times as high," Walters said. He ran to get a Polaroid camera and took a picture, then three more, from a distance of about 150 feet.

The camera was now out of film, so Walters dashed inside, reloaded, and took another picture as the UFO drifted closer. He was about to snap yet another picture, he related, when a blue beam shot from the object, lifting him off the ground and immobilizing his muscles. There was a sickening stench, a sort of ammonia-like smell with overtones of cinnamon, and Walters heard what he described as a "computer-like voice" inside his head. It said, "We will not harm you." Walters remembers screaming. Then he fell. Looking up, he saw that the UFO had vanished.

That was just the beginning. Nine days later, at 4:00 in the afternoon, Walters returned home and began to hear a humming sound. But he saw nothing and pushed it from his mind, until the hum changed to a mechanical-sounding voice accompanied by what seemed to be a rush of air inside his head. Walters grabbed his camera and went outside. "I hear you, you bastards," he cried out, and the voice responded, "Be calm. Step forward." At that point, Walters saw a rapidly descending dot of light in the sky and raised his camera only to hear a voice order him not to take the picture. *"Las fotos son prohibido,"* it said in Spanish. But Walters snapped a picture of the UFO hovering over nearby houses. Other voices urged him to come on board the UFO. A woman's voice filled his head: "They won't hurt you. Just a few tests that's all. They haven't hurt us and we are going home now." Walters refused to cooperate and yelled up that they had no right. "We have the right," came back a male voice, and then it said, "We will come for you." Walters shot another picture just as the UFO disappeared.

The Walters family tried to maintain a normal life in spite of the disturbing intrusions. As soon as the UFO departed that night, Ed rushed off to the high school homecoming football game, where his son, a drummer in the band, appeared in the halftime show. But life had irrevocably changed. Over the next few days Ed and Frances talked of little else but UFOs every night and when they were alone in the daytime. Ed loaded his .22-caliber rifle and his .32-caliber pistol. And he started regularly carrying a shotgun behind the seat of his pickup.

Then, at 3:00 a.m. on December 2, Walters awoke to voices in his head. He grabbed the pistol and with his wife went to the kitchen window. They saw the UFO, diving down to hover about 100 feet over the swimming pool. "Step forward now," directed a voice. Instead, with the camera in his right hand and the firearm in his left, Walters slipped from the house, took another photo, and ran back inside. Peeking out, he and his wife saw the UFO move off.

Incredibly, there was more that night. Walters said he and his wife had just gone back to bed when the dog barked, and Walters, pulling open the curtains of a French door, saw a "creature" standing outside. Walters described it as four feet tall and wearing a helmet through the faceplate of which Walters could see a pair of large black eyes; a sort of box covered the thing's torso and another smaller box enclosed it from hips to knees. In its hand it carried a shining silver rod.

The terrified Walters let out a yell, then tripped and fell backward. Frances crawled over the bed and saw the creature too. Ed grabbed his pistol to shoot the intruder if it attempted to enter. But it simply stood outside regarding him with eyes, he said, "that were almost sad. Eyes that somehow seemed curious."

Gathering his courage, Walters thought that maybe he could capture the little being. But as he opened the French doors, the blue beam hit his right leg, paralyzing it, he said, as if it were "nailed to the floor." Then the beam rose and started to lift his leg, and Walters saw the UFO hovering fifty feet overhead. Desperately, he grabbed the doorjamb, and Frances clung to him from behind. Suddenly, his leg was free. The beam retreated and the UFO vanished. But not before Walters had snatched up his camera and gamely shot yet another Polaroid picture—his eleventh.

There were five more sightings that month of December with another handful of Polaroids. When the UFO appeared on the evening of the twenty-eighth, Walters had no more film for the Polaroid, so he took his Sony Camcorder and shot a one-minute, thirty-eight-second videotape of the object passing behind a tree in the backyard. His son, Dan, and daughter, Laura, both said they saw the UFO and witnessed their father taping it. Later, this would turn out to be one of the most significant records of all.

On into the New Year the siege continued, with five episodes in January. The beings now started to call Walters by the name of "Zehaas." "Come forward. We will not harm you," a voice commanded on January 12. "Zehaas, we are here for you," the voice announced on January 26, adding mysteriously, "In sleep you know." Walters's response was more photographs of the UFOs. He took to shouting obscenities at the objects and yelling "Get the hell out of my life." But the voices kept repeating, "We are here for you, Zehaas. Do not deny us. We are here. Remember."

By now, Walters had confided in Duane Cook, editor

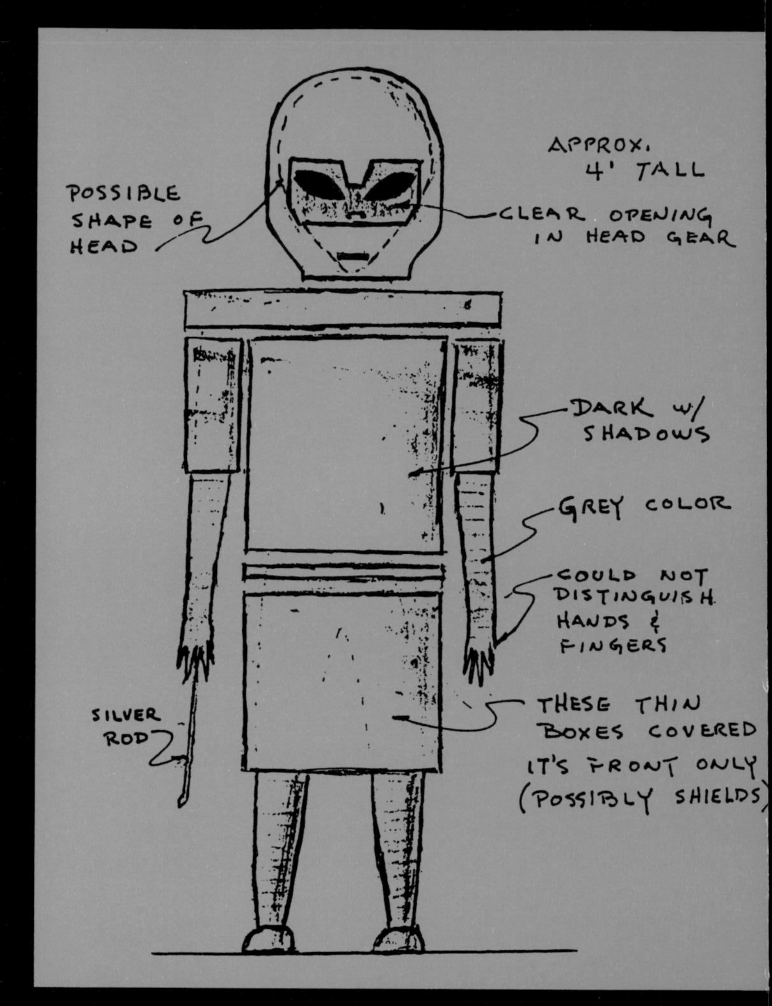

The drawing above, by Edward Walters, depicts the "small, shielded being" that he claims appeared outside his bedroom window in the predawn hours of December 2, 1987. Walters was struck by the creature's strange black eyes, which to him "resembled the contoured eyes on a grasshopper." When the alien moved away and Walters tried to pursue it, he was hit by a blue beam from a hovering UFO, which froze him in place. After struggling out of the grip of the beam, Walters says, he photographed the UFO as it flashed down another beam (large picture, right), which he believes transported the alien to the spaceship. According to Walters, about two months later his wife, Frances, was outside when she narrowly escaped being captured and drawn up by a similar beam—which he somehow once again managed to catch on film (inset, right). Treated by a process called light-blasting, the photograph reportedly shows Frances dodging the beam as she runs into the house for cover.

ANIMAL EFFECT "DURING" UFO ENCOUNTER

(Please elaborate on items checked below, where applicable, by using a separate

AUDIBLE: BARKED ( ) HOWLED ( ) WHINED ( ) BELLOWED ( ) OTHER______

EMOTIONAL: INDIFFERENT ( ) STARTLED ( ) PANICKED ( ) HID ( ) OTHER______

MENTAL: DISOBEDIENT ( ) BEWILDERED ( ) OTHER______

MOVEMENT: STILL ( ) COWERED ( ) SHOOK ( ) PACED BACK AND FORTH ( ) PACED IN A C

KICKED ( ) COVERED EARS ( ) FLEW ( ) OTHER______

PHYSICAL: PARALYZED ( ) BLINDED ( ) BURNED ( ) URINATED ( ) HAIR STOOD ON END

LOSS OF BALANCE ( ) LOST CONSCIOUSNESS ( ) DIED ( ) OTHER______

COMMENTS: ______

of the Gulf Breeze *Sentinel.* Cook, while withholding Walters's last name, identifying him only as "Mr. Ed" or "Ed," ran stories about the incidents and published some of Walters's pictures. Cook also joined Walters on occasion in hopes of being present when a UFO arrived. The two men were riding together in Walters's pickup on January 24, when Walters heard the dread humming sound and the voice reminding him, "In sleep you know." Halting the truck, he stepped out onto the road but could see nothing. Cook, meanwhile, backed off a distance and started filming Walters. As both men were climbing back into the truck, the UFO loomed overhead, and Walters snapped a photo. By the time Cook turned around, the UFO was gone, but Walters handed the Polaroid to him, and when Cook pulled the picture from the camera, there was the image of a UFO.

For more than a decade, people in the Gulf Breeze area had occasionally reported seeing UFOs. When word of Ed Walters's sightings got around, accounts of other incidents proliferated, and many of them turned out to involve the same dates and vicinities as Ed's reports. On November 11, 1987, for instance, the night Walters said he first saw a UFO, at least seven other people claimed sightings of their own, some fairly similar to what he described. One reported a blue beam.

Newspaper and television coverage attracted the attention of UFO investigators to Gulf Breeze and in particular to Ed Walters. Representatives of J. Allen Hynek's Center for UFO Studies (CUFOS), the Mutual UFO Network (MUFON), and the Fund for UFO Research (FUFOR), as well as local ufologists, interviewed Walters and looked at his pictures. MUFON investigators supplied Walters with a special four-lens Nimslo 3-D camera that would provide four independent negatives of each picture, three of which could be subjected to various analytical tests while one was preserved intact. The slight differences in the positioning of the four lenses also would provide a basis for calculating size and distance of a photographed object. But the Nimslo camera had an even more important feature. It was tamperproof.

Inevitably, critics had begun raising doubts about Ed Walters's Polaroid pictures, questioning whether they might be the result of trickery, perhaps some clever method of double exposure. The Nimslo camera was loaded with film and sealed with wax by two MUFON representatives before it was given to Walters. Moreover, the MUFON team shot an identifying photo on the film in the Nimslo; if, despite the seal, someone managed to doctor the film or substitute a new roll, the damage to the identification image, or its absence, would reveal the tampering. Finally, if and when Walters managed to get a shot of a UFO with the Nimslo, the camera would be unsealed in the presence of journalists at a press conference, and at least two witnesses would stay with the film all the way through the developing and printing processes to guard against fraud.

Walters realized soon after he accepted the Nimslo that it put him in a dilemma. "If I did not take a picture with

Tools of the ufologist's trade, detailed questionnaires such as these published by the Mutual UFO Network (MUFON) enable field investigators to collect data on reported sightings in a consistent and thorough manner. The MUFON forms cover everything from reactions of animals to UFOs to descriptions of alien beings. The form excerpted at right is used to question people who have encountered UFOs on radar.

HEAD AREA (Shape/No. of): HEAD ______ DESCRIPTION OF ENTIT
MOUTH ______/# ____ EYES ____
HEAD COVER (Similar to): HAT? ( ) HOOD? ( ) HELMET? ( ) HAI
/# ____ EARS ____
DY AREA (Specify): TORSO? ( ) ARMS? ( ) HANDS? ( ) FINGE
FEET? ( ) GLOWED? ( ) NO./FINGERS-CLAWS
COVER (Similar to): COVERALL? ( ) WETSUIT? ( ) DIVING SUIT? (
GLOVES? ( ) SHOES? ( ) BOOTS? ( ) OTHER
Miscellaneous): SKIN COLOR ______ SKIN TEXTURE ______
HAIR COLOR ______ HAIR LENGTH ______
SUIT COLOR ______ SUIT TEXTURE ______
Y FELT TO TOUCH: ______
COMMENTS: ______

DID THE ECHO(s): (Please elaborate upon items checked below by using a separate sheet of paper)
APPEAR ABRUPTLY? ( ) MANEUVER? ( ) MOVE IN A FIXED FORMATION? ( )
APPEAR GRADUALLY? ( ) CHANGE IN QUALITY? ( ) APPEAR VISUALLY WHILE TRACKING? ( )
DISAPPEAR ABRUPTLY? ( ) CHANGE IN SHAPE? ( ) PACE A KNOWN TARGET? ( )
DISAPPEAR GRADUALLY? ( ) CHANGE IN SIZE? ( ) DIVIDE INTO MULTIPLE TARGETS? ( )

WERE ANY OF THE FOLLOWING NOTED? (Please elaborate upon items circled below on a separate sheet of paper)
(INDICATE "WHEN" NOTED BY CIRCLING WHETHER OBSERVED - PRIOR, DURING, OR AFTER INITIAL OBSERVATION)
DUCTING? PRIOR DURING AFTER AIRCRAFT? PRIOR DURING AFTER
SUPER-REFRACTION? PRIOR DURING AFTER BIRDS? PRIOR DURING AFTER
SUB-REFRACTION? PRIOR DURING AFTER WEATHER? PRIOR DURING AFTER
GROUND CLUTTER? PRIOR DURING AFTER INTERFERENCE? PRIOR DURING AFTER
ANGELS? PRIOR DURING AFTER RADAR MALFUNCTION? PRIOR DURING AFTER
OTHER? ______ PRIOR DURING AF

WAS THERE OTHER CONFIRMATION OF THE UFO(s)' PRESENCE? (INDICATE BY ENCIRCLING "WHEN" BELOW AS APPLICABLE)
RADAR? PRIOR DURING AFTER VISUAL? PRIOR DURING AFTER
DESCRIBE: ______

WHAT WERE THE UFO TARGET(s)' POSITION RELATIVE TO RADAR SHADOW AREAS DURING THE OBSERVATION? ______

EQUIPMENT TYPE/OPERATION
RADAR TYPE: DOPPLER ( ) CW ( ) PULSE ( ) MTI ( ) OTHER ______

this special camera, it would reflect on the credibility of the case," he said. "I had been hoping the UFO would get out of my life. Now I had to pray that it would return for at least one more photograph."

For two weeks Ed and Frances Walters watched the night skies in hopes of capturing the UFO's image with the Nimslo. At last, on February 26, 1988, Ed Walters related, they sighted a strange object over the waters of Santa Rosa Sound on the south side of Gulf Breeze. But it looked completely unlike the UFOs Ed had been reporting and taking pictures of up till now. This one had an elongated, cigarlike shape, three longitudinal rows of orange lights, and a trailing light, and it created a visible "peculiar atmospheric disturbance," he said, like heat waves shimmering over a hot asphalt pavement. He shot ten exposures of the object, insisting to his wife all the while that the thing appeared to be huge and a long way away. Frances Walters disagreed; she thought it seemed to be relatively small and not all that far from where they stood.

When the camera was publicly opened and the film processed a week later, the images at first led the MUFON researchers to agree with Ed that the pictured UFO was very large. In fact, they began referring to it as the "mother ship," apparently implying a relationship to the smaller craft in the Polaroid photos. But careful analysis of the Nimslo negatives later showed that Frances Walters had been much nearer the mark. The object in the pictures was only about three feet long. Now the investigators dropped the mother-ship appellation and instead dubbed it a "probe." Despite this demotion, to all appearances it remained a UFO. It was definitely not an airplane, nor would it yield to any other conventional explanation. And it had been photographed with what investigators considered a tamperproof camera.

In February, a skeptical scientist, Bruce Maccabee, came to Gulf Breeze to look into Ed Walters's claims. Maccabee was the chairman of FUFOR and a respected optical physicist working for the U.S. Navy at a research establishment in Maryland. "I suspected a hoax," Maccabee said, and he had thought he could expose it in three days. But after interrogating Walters and other witnesses for hours, visiting the sites, and conducting an initial analysis of the photos, he realized the matter was going to require more study—and he was no longer convinced of the outcome.

Not long after leaving Gulf Breeze, luggage crammed with pictures, tape recordings, and other evidence, Maccabee suggested in a phone conversation that Walters build a "self-referencing stereo" (SRS) camera. He could do this by fixing two Polaroid cameras about a foot apart on a board and positioning a pole to stick out in front between them with a vertical nail on its tip. If the cameras were shot simultaneously, the relative positions of the pole tip and the photographed object in the pictures would provide information to calculate—at least roughly—the distance of the object from the cameras, and thus the approximate size of the object. In other words, with pictures from this device it would be easy to tell a hoaxer's model from a large craft.

"As I explained how to build this SRS camera," Maccabee recalled later, "I wondered what his response would be; if this were all a giant hoax, I was making a suggestion that could expose the hoax." Walters's response was to build the SRS camera, to improve it based on follow-up suggestions from Maccabee, and to shoot a number of pictures of alleged UFOs with it. Taken together with the interviews, the earlier Polaroids, the Nimslo pictures, and the videotape, the photos from the SRS device convinced Bruce Maccabee that Ed Walters—and his pictures—were for real.

The final episode in Walters's travail took place at 1:15 on the morning of May 1. He had gone to the town park in hopes of securing still more photographs when he sensed the familiar humming sound. Two UFOs appeared and while Walters was photographing them, his vision, he said, went "completely white just as if a flash cube had gone off in my brain." He lost all sensation in his body but felt that he was falling. When he came to, his watch showed 2:25 a.m. There was a large bruise surrounding a red dot on the bridge of his nose and two similar bruises at his temples. The fingernails of his right hand were filthy with a black, evil-smelling substance. Ed Walters had no memory of the lost seventy minutes. But it seemed reasonable to think that perhaps the aliens had come at last for Zehaas.

Ed Walters and his sensational accounts have from the start stirred wide debate and major controversy among ufologists. Researchers seek to establish a norm, a consistency in UFO descriptions that would argue for a true phenomenon and not just a random collection of misperceptions, fantasies, and outright lies. Anything far outside the pattern is generally considered suspect, and in fact many cases in which subjects allegedly had repeated close encounters or offered numerous photographs have turned out to be the work of hoaxers or strongly suspected hoaxers.

Yet neither Ed Walters nor his family fit the profile of the classic UFO hoaxer. Though he and his wife eventually did write a book about their experiences, they were in no need of money and were known to the community as quite the opposite of jokers or publicity seekers. Walters had insisted on anonymity for more than a year because he feared that publicity would actually hurt his construction business. Moreover, in February 1987, at the height of the purported UFO assault on him, Walters willingly underwent a pair of polygraph tests that covered his photographs as well as his visual recollections. He passed both tests without difficulty. And while it is true that certain sociopathic personalities can on occasion defeat a lie detector, Walters in June gladly submitted to a thorough evaluation by a prominent Florida clinical psychologist. The psychologist found no evidence of any mental disorder.

Perhaps the strongest support of Ed Walters and his Gulf Breeze sightings comes from FUFOR's Bruce Maccabee. An expert photoanalyst as well as an energetic and experienced investigator of UFO sighting claims, Maccabee devoted hundreds of hours over more than a year to his study of the Walters case.

From the outset, he considered the possibility of multiple exposures, which he knew could be accomplished even on a Polaroid. Basically, this would involve shooting a model against a black background and then exposing the same film to an exterior scene in such a way that the image of the model would appear against a dark night sky. There were problems and pitfalls involved, however, which could leave telltale clues in the pictures even if they were created by an expert photographer—and Maccabee quickly became convinced that Walters was no expert. "The evidence seemed almost overwhelming that Ed had not perpetrated a hoax," he concluded. "The sightings really happened."

The optical physicist focused his most painstaking examination on the ninety-eight-second videotape shot by Walters that purportedly showed a UFO passing slowly behind his house on December 28, 1987. Maccabee put the tape through a rigorous computer evaluation for size of the object, distance from the camera, and lighting variations, as well as speed and track. He reproduced—with Walters's assistance—various hoax hypotheses: that the UFO was a model hanging from some sort of support, that it was made to move by a rope-pulley-rail arrangement or was carried atop a pole. Maccabee judged that none of the hoax possibilities could account for the videotape. "The overall image shape appears to be identical to the image shapes in Ed's Polaroid photos of November and December, 1987," Maccabee wrote at the end of a lengthy paper. "I conclude that Ed did not produce his videotape using a model. Rather, I conclude that he videotaped a true UFO."

ebunkers, however, have remained just as firmly convinced that the whole affair is an elaborate hoax and have launched recurrent public attacks on Walters's credibility. Among the most damaging was a series of three challenges that arose in quick succession in 1990. First, a model of a UFO made from styrofoam plates allegedly turned up in the attic of a house Walters had sold in 1989. Then an enterprising local television reporter, Mark Curtis, demonstrated on the air how he achieved still photos very much like Walters's by using models of UFOs and double exposures. He also produced a videotape of a cruising "UFO" by shooting a lighted model carried around on top of a tall black pole at night. At about the same time, a young man named Tommy Smith, a friend of Ed Walters's teenage son, came forward and publicly declared that he had been present on occasions when Ed gleefully boasted how he was deceiving the world with models and fake photos. "He thought it was very funny," Smith said of one incident Walters described. "He rolled on the floor laughing."

Walters and his supporters scornfully rejected all the charges. The model that supposedly was found in his former house, Walters said, was certainly not put there by him. Moreover, he said, it was "a kid's paper plate model, a blatantly different thing" than the UFOs he saw and photographed. The television reporter's photos were also said to be very different from Walters's pictures—and, anyway, Bruce Maccabee had acknowledged from the outset of his investigation that double exposures were possible; the essential point was that careful study had convinced Maccabee that Ed Walters had not faked his photos. As for Tommy Smith's allegations, it seemed to be his word against Walters's.

The sniping and counterfire continued, dividing much of the ufology community into warring camps. Walters says the debunkers have "tried to smear me." One anonymous detractor, he said, even doctored a picture of the Chrysler Building—adding a crude, obviously fake image of a UFO circling the top of the edifice—and sent it to a newspaper as a photograph by Walters. A more serious attack came in early 1992, when an independent engineering consultant and photo specialist named William G. Hyzer completed a study he had been asked to undertake by some concerned ufologists. Hyzer, characterized as completely neutral by those who enlisted him, criticized a number of Walters's Polaroid shots as suspect. But in only one case did he make an out-and-out accusation of fraud.

The picture in question, "Photo 19" by Walters's own numbering system, was shot January 12, 1988, and shows an image of a UFO hovering over a highway in front of Ed Walters's Ford 150 XLT truck. A strong light from the "power ring" at the bottom of the alleged craft is reflected on the roadway below. The hood of the truck, visible in the foreground of the picture, does not reflect any of this light. After conducting tests with another Ford 150 XLT and a light positioned at various points over the highway, Hyzer decided that if Walters's picture were genuine, the Ford hood would have shown a reflection. "It is this author's professional opinion," Hyzer wrote, "that there is only one logical explanation for this optical anomaly: photograph number 19 is a fake produced by multiple exposure photography."

Although some of Walters's critics seized on this pronouncement as the proof positive they were seeking, Walters maintains it does not injure his case at all. There was no reflection on his truck hood, he said, because the vehicle had "been in a wreck; the hood had been bent." Maccabee backs him up. As long ago as 1988, the physicist said, he himself had dealt with that specific point, conducting extensive tests with Walters's own truck to determine whether a reflection would appear on the hood in the situation depicted in Photo 19. Because the shape of the hood had been distorted when a backhoe reversed into it some time before, no reflection would show, Maccabee found, unless the light was about nine feet above the road—considerably higher than the light in Photo 19.

The Gulf Breeze story did not end with Ed Walters's purported missing-time experience in May of 1988. Reports of UFO sightings in the vicinity continued. Indeed, they came so fast and furiously that in November of 1990 a team of seventeen investigators working to MUFON guidelines began keeping a formal nightly watch for UFOs over the Gulf Breeze area. By March of 1992, the investigators had recorded an astonishing 130 sightings, documented by some two hours of videotape and dozens of still photos.

Most sightings involved distant lights that turned from red to white and back again. Sometimes the lights were in the form of a ring. Local military bases were said to be unhelpful when queried about air traffic, but the watchers became convinced the lights, which moved noiselessly, were not any kind of known aircraft. Flares hanging from balloons were suggested as an explanation, but the lights often moved directly counter to the wind, and they turned on and off in a blink of an eye; how could a hoaxer ignite and extinguish flares in the sky? And "diffraction spectral analysis," said one investigator, revealed the glow to be very different from that produced by flares.

On some occasions, the object was not so ill-defined. In September of 1991, more than eighty thrilled witnesses reported a dark disklike shape that moved silently across the sky and suddenly became illuminated, as if by lights behind porthole windows. "It was the same UFO that Ed Walters photographed three years ago," declared one observer. Two months later, on November 15, 1991, more than a dozen watchers saw a similar display, this time involving not only "portholes" but also a "power ring" below.

A number of active and retired military people have been among the witnesses in Gulf Breeze. "They say they can't conceive of what it might be," said Bruce Maccabee, who himself witnessed one appearance of the UFO. Mac-

Ed Walters's famous "Photo 19" (below) is among the most controversial of his Polaroid shots of purported UFOs over Gulf Breeze, Florida. Here light-enhanced to reveal more detail, the picture was condemned as a multiple exposure fake by one respected photoanalyst, William G. Hyzer, and endorsed as authentic by another, optical physicist Bruce Maccabee. A resourceful television news reporter, Mark Curtis, faked the similar photo opposite by multiple exposure to prove it could be done.

cabee, too, admitted being puzzled about whatever was haunting the skies over Gulf Breeze. "It's either a hoax," he said, "or the real thing—whatever that means."

In addition to photographs, there is another kind of "proof" that UFO investigators examine with great care, and that is purported physical evidence of a UFO's visit. Such "traces," as ufologists call them, are relatively uncommon considering the number of UFO reports. Yet the data banks nevertheless contain thousands of examples, and to some investigators they offer strong evidence of extraterrestrial visitation. The difficulty lies in establishing authenticity, for any number of things can leave exceedingly odd traces.

A physical phenomenon that lately has attracted increasing attention from ufologists—as well as from researchers with other special fields of interest—is crop circles, those puzzling, often complex, patterns of flattened and swirled wheat or other grain plants that have appeared inexplicably on farms all over the world. Some ufologists believe the circles may be traces of alien activity. Most attention has been focused on British examples, since they have seemed to be most numerous—appearing at the rate of hundreds a year—and often spectacular in dimension and design. Thus it was that when a pair of British artists in their sixties claimed in 1991 that they had created the circles as a prank, much of the superficially informed public accepted that the mystery had been solved.

Investigators of crop circles, however, were not long in dismissing the assertion of the two men. It was apparent that the pair did not have time or opportunity to make even all the circles in Britain, much less those in the rest of the world. Nor could the techniques they described—dragging planks while guided by stretched strings or sighting devices—account for the characteristics of many circles: stalks laid flat but still alive and unbroken, narrow bands flattened in opposite directions around the circle, intricate braidlike patterns of bunches of plants alternately laid over and under the bunches next to them, and rings precisely shaped on steep hillsides where footing is difficult to maintain.

Serious crop circle researchers—they call themselves cereologists—have turned up other information that lays waste to the hoax explanation. Marshall Dudley, for instance, a systems engineer at Tennelec/Nucleus in Oak Ridge, Tennessee, used highly accurate radiation measuring devices to compare soil samples from several crop circles with control samples collected outside the circles. What he found was both fascinating and puzzling. Some soil from circles gave off considerably less alpha and beta radiation than their controls, while samples from other circles emitted significantly higher levels of radiation than soil collected nearby. The alpha and beta elevations were steep enough to occasion a health caution from Dudley: Anyone who actually sees a crop circle form should stay out of the immediate area for several hours.

Dr. W. C. Levengood, a Michigan biophysicist, compared wheat plants from cir-

cles with control plants taken from the same fields but outside the circles. He found that the growth nodes in the stalks of the circle plants were swollen, as if they had been heated or otherwise subjected to a quick burst of intense energy. Under microscopic examination the cell walls of the circle plants were seen to be stretched and distorted, again as if heated rapidly, enlarging the tiny apertures through which ions and electrolytes pass into and out of cells. And up to 40 percent of the seeds from the circle plants were malformed, versus no malformed seeds in the control wheat. The same tests were run on plants taken from crop circles known to be created by hoaxers; none of the anomalous effects were present.

Researchers like Dudley and Levengood do not necessarily look to UFOs as an explanation for crop circles, but their work is of great interest to ufologists who do make a connection. And so are the many field studies of crop circles during which investigators have observed strange lights and heard weird noises, much like those reported by people who say they have seen UFOs.

Nothing excites ufologists as much as physical traces associated with a full-fledged allegation of a UFO sighting. Such was the case in Delphos, Kansas, a small prairie town devoted to wheat and livestock, according to the following account related by a sixteen-year-old farmboy named Ronald Johnson and his parents. The young Johnson said that at dusk on November 2, 1971, he was startled to see an illuminated object hovering near a tree seventy-five feet away. The thing seemed to be about nine feet in diameter and ten feet high; it glowed red, orange, and blue, while giving off a rumbling noise that Ronald later described as "like an old washing machine that vibrates." The UFO seemed to be suspended about two feet off the ground, which was lit by a bright glare emanating from the object's underside. He said the light was so intense that it hurt his eyes.

The teenager stood transfixed for some minutes. Then, the UFO started to move off, and Ronald ran to get his parents. They saw the object receding into the distance, and where it had hovered over the ground there remained a brightly glowing ring of soil. Mrs. Johnson hurried to the house for a Polaroid camera and photographed the luminescent ring. The Johnsons then touched the ring. It felt cool, but they immediately experienced a numbing effect in their fingers, something akin to a local anesthetic. They described the soil as having "a slick, crust-like" texture, as though "crystalized," and it seemed to be blistered.

The next day, the sheriff came out to collect soil samples and photograph the ring, which had stopped glowing by then. A month later, MUFON investigator Ted Phillips, Jr., arrived in Delphos to interview the Johnsons and examine the site. It had snowed and thawed, and the ring exhibited an extremely odd characteristic—a hydrophobicity, or aversion to water, that left it covered with unmelted snow in contrast to the muddy earth all around. This hydrophobicity lasted all through the winter.

In the course of the investigation, both photographs proved inconclusive. On analysis, Mrs. Johnson's Polaroid showed considerable reflected moonlight that could be taken for an internal glow; the sheriff's photo, shot nineteen hours after the event, showed a whitish surface to the earth that would reflect moonlight. Nevertheless, researcher Phillips was highly impressed with the credibility of the Johnson family and sent off soil samples to Erol A. Faruk, an English soil analyst.

In his published report, Faruk concluded that while deception was possible, the presence in the soil of an unidentified compound with "unusual characteristics" made a hoax "the least plausible of explanations." The puzzling compound appeared to be a highly unstable silver salt, which, when exposed to air, would oxidize to produce a fluorescent substance.

Faruk speculated that a "hovering object of presently unknown origin" possibly did appear over the Johnson farm that November evening. He argued that it could have "contained within its periphery an aqueous solution of an unstable compound whose sole function would be for light emission"; that "some of the solution was deposited into the ground" while the object hovered low; and that "the rum-

Often attributed to UFOs, some crop circles—such as this elaborate design discovered in 1991 on Fred Watmough's wheat field in Lethbridge, Alberta, Canada—reportedly cause headaches and nausea in people who enter them. Animal behavior, too, is said to be affected: Dogs display agitation just before patterns are found. And one investigator noted that geese avoided flying over Watmough's circles.

Photomicrographs of wheat plants taken from crop circles on a British farm show obvious swelling in the cell walls (lower left) compared with those of normal plants (top left). Pits in the cell walls—the dark dots in the lower picture—also expanded significantly; normally they are difficult to photograph. Instead of healthy wheat seeds (top right), another British crop circle produced seedless heads (lower right) that display a rare mutation called polyembryony—multiple embryos from which seeds never develop. In both cases the damage was judged to be caused by high heat or other radiation.

bling noise heard at the same time might be associated with the manner in which the deposition occurred. Once enough of this essentially expendable solution was ejected the object departed while Mr. and Mrs. Johnson approached the ring area."

Faruk offered no suggestions as to what the reason for such an event might have been. He simply said his research indicated that it could have happened just as the Johnsons described. Further study "should provide a conclusive answer, not the least of which is the identity of the soil compound," Faruk wrote.

A frustrating silence has been the stance of the authorities and most government-connected scientists for the past two decades and more. In 1969, a civilian scientific committee under contract to the United States Air Force and chaired by the distinguished physicist Edward U. Condon issued a report saying that nothing had come of the Air Force's twenty-two-year study of UFOs. Condon firmly advised that further study was a waste of time, and later that year, the air force shut down Project Blue Book, its UFO information-gathering agency. Thereafter no U.S. government agency maintained an active UFO research program, although in the 1970s the National Air and Space Administration took on the task of answering letters about UFOs addressed to the White House or to itself.

Among the propositions thus cast into official oblivion was the intriguing idea that one or more UFOs had crashed and that the federal government had retrieved both wreckage and bodies. Some accounts put the number of recoveries at forty—with no fewer than nine saucers stored in one hangar at a secret Nevada base. American pilots, it was whispered, had even flight-tested an operable spacecraft. Other stories described dozens of alien bodies in U.S. military morgues, and more astounding still, that a few aliens had survived to become "guests" of air force intelligence. These aliens, the rumors buzzed, were helping U.S. designers build vehicles with antimatter propulsion systems so advanced that by comparison the 2,500-miles-per-hour SR-71 Blackbird spy plane would seem as primitive as Leonardo da Vinci's first parachute design.

Most of the stories were obvious fables, the work of superheated imaginations. But perhaps not all, and one crash-retrieval report that long ago had been convincingly shot down by the military—the famous Roswell, New Mexico, case—seems to have been brought back to life in recent years by infusions of new evidence and testimony. On the morning of July 8, 1947, the public relations officer at an air base in southeastern New Mexico issued an electrifying press release: "The many rumors regarding the flying disks became a reality yesterday when the intelligence office of the 509th Bomb Group of the Eighth Air Force, Roswell Army Air Field, was fortunate enough to gain possession of a disk through the cooperation of one of the local ranchers and the sheriff's office of Chaves County." The press release also said that the alien object had crash-landed in the desert a day or two before and that air force technicians at Roswell had examined the wreckage before sending it on to "higher headquarters."

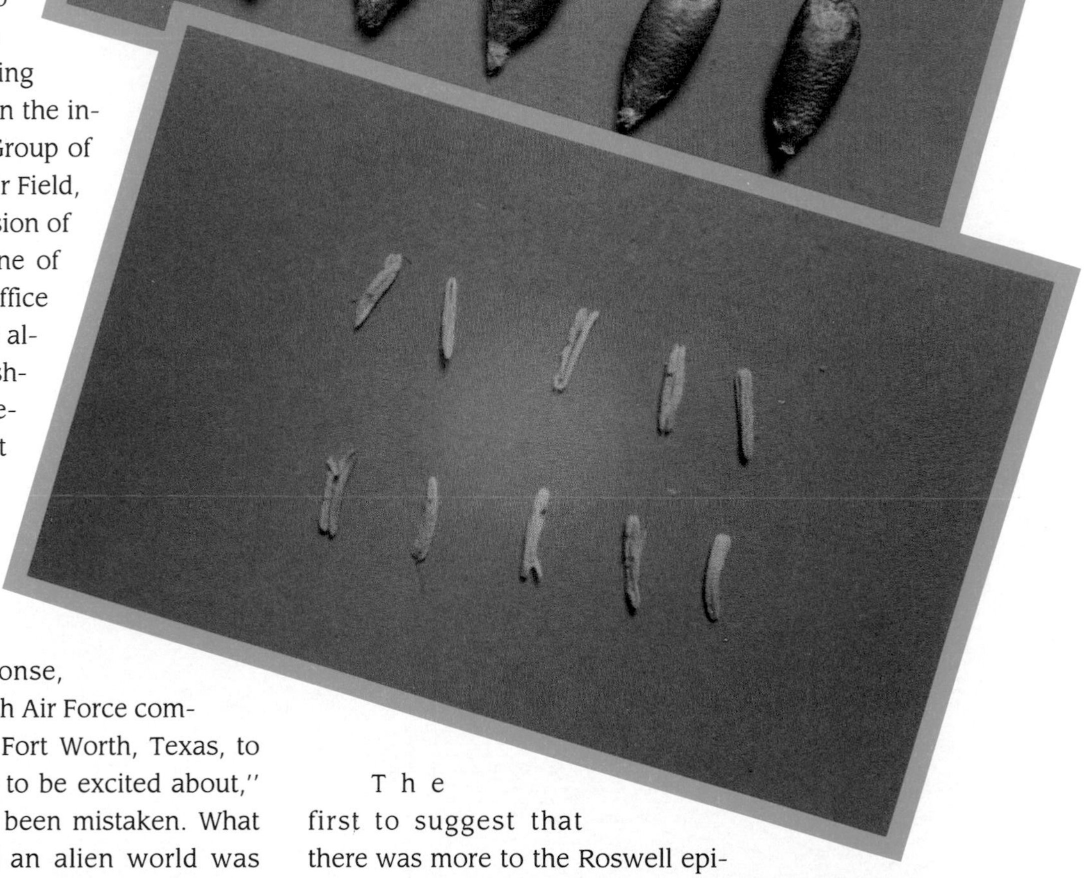

The Roswell switchboard soon erupted with calls from reporters all over the world. In response, Brigadier General Roger Ramey, Eighth Air Force commander, held a press conference in Fort Worth, Texas, to set matters straight. It was "nothing to be excited about," he said. The officers at Roswell had been mistaken. What they thought to be a vehicle from an alien world was merely a remnant of a new type of radiosonde weather balloon with a radar reflector. The general displayed the debris, while an air force weather officer stepped forward to brief reporters further and photographers were invited to snap pictures.

"General Ramey Empties Roswell Saucer," snorted one headline. The papers reported that Washington had handed the Roswell officers "a blistering rebuke" for their sensation seeking. That seemed to be that; the episode was shelved. And there it remained for almost thirty years—until witnesses started coming forward to tell the press and UFO researchers quite a different story.

The first to suggest that there was more to the Roswell episode was Jesse A. Marcel, an army major and base intelligence officer at Roswell in July of 1947. After years of silence, Marcel in 1978 apparently decided it was time to tell his story and did so in interviews with several investigators. What he had to say helped lead researchers to others with firsthand knowledge of the case—or in the instance of some who had died, their surviving children.

Marcel said he had been one of the officers who raced out to a ranch northwest of town upon receiving word of a thunderous explosion and a field filled with wreckage. What they found was a 500-foot-long gouge in the earth and a 3,500-by-300 foot area thickly littered with a dull gray

Highlighting the red color components in this reproduction of the original photo at center obscures the body of the craft.

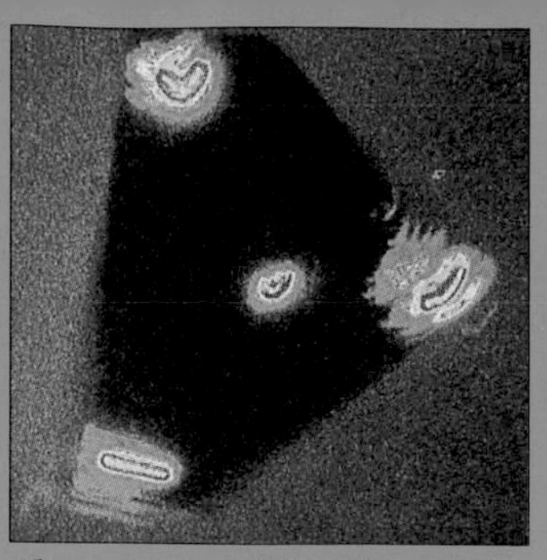

The contours of the triangular object are most clearly defined by the blue false-color version of the original photograph.

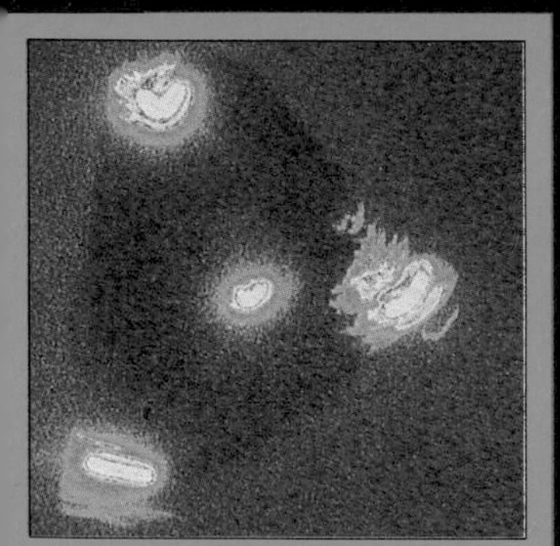

This false-color reproduction emphasizes the green components of the photograph, revealing a shadowy triangle.

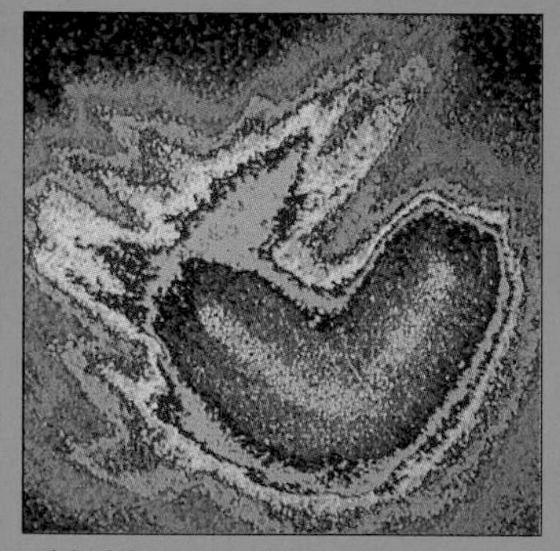

This blowup of the top corner of the strange object, reproduced in false color, affords a very detailed view of the light itself.

Official Confirmation of an Amazing UFO

In the spring of 1990, Belgium won a place in the annals of ufology when its government became the first to involve itself in an open UFO inquiry. Since the previous November, several thousand residents of the country's Wallonia region had reported seeing a strange triangular craft with three brilliant lights moving slowly in the night sky. On the evening of March 30, 1990, Belgian Air Force radar picked up the object as well. The air force scrambled two F-16 fighter jets to investigate.

The jets' computerized radar soon

peared as a diamond shape on their screens *(far right).* But within seconds after the radar fix, the craft accelerated away with astonishing speed—so fast, say the experts, that the G forces would have crushed a human—then dove until it eluded the radar. The chase was twice repeated before the jets were ordered back to base.

Besides its unprecedented release of data about the incident, the government took another unheard-of step: It cooperated with a private UFO group, the Belgian Society for the Study of

affair. Witnesses' photographs of the UFOs were analyzed at the country's Royal Military Academy. The center photograph above, for instance, taken in April 1990, was reproduced in so-called false colors as seen in the surrounding pictures to provide clearer definition of the object; the academy found no evidence of a hoax.

Skeptics suggest that the sightings, which began to taper off by August 1990, might be attributed to the testing of secret military aircraft—but that explanation was dismissed by the

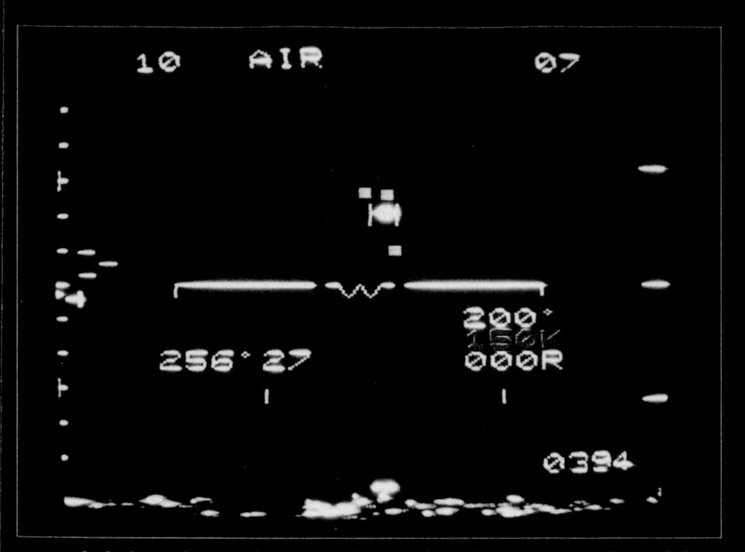

Recorded aboard one of the F-16 fighter jets that pursued the Belgian UFO, this sequence of three images of the pilot's radar screen shows the movement of the target—which appears as a diamond—over an interval of just one second. The UFO's speed in nautical miles per hour, or knots, is seen in red numbers. The number at the top right of the screen denotes the object's altitude in thousands of feet.

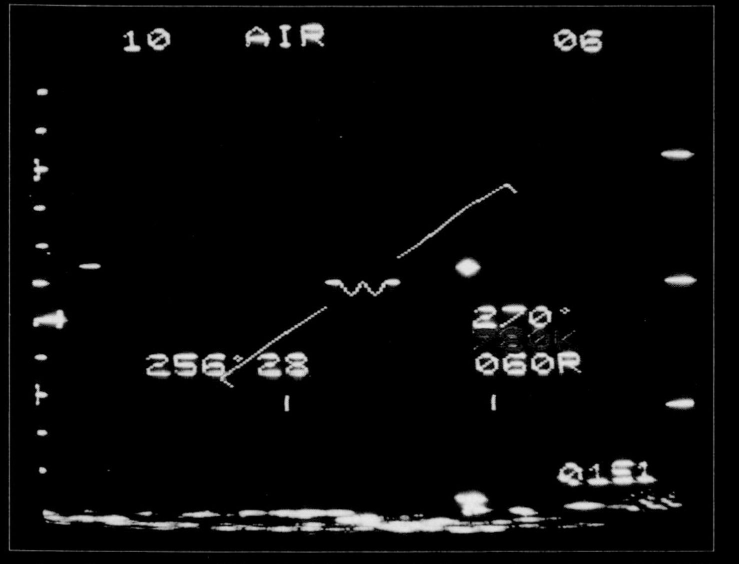

In less than a second, the UFO dove 1,000 feet and accelerated to 780 knots. This should have created a sonic boom—but none occurred.

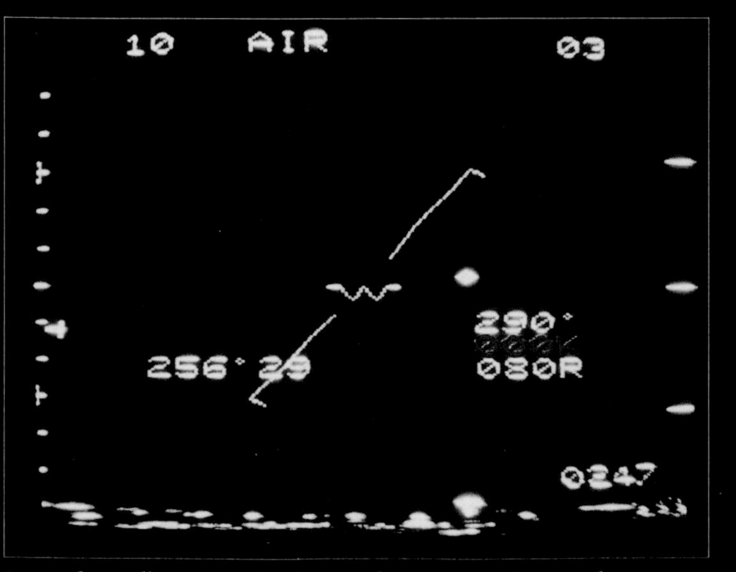

In another split second, the UFO sped up to 1,000 knots—shown as 000, since the F-16's radar does not record speeds above three digits.

metallic detritus: fragments of thin, featherlight but virtually indestructible material that could not be bent, dented, cut, or burned; other bits of a foil-like substance that when crumpled would unfold itself with nary a wrinkle; chunks of what looked like I-beams, also exceptionally light in weight and one of them embossed with a curious sort of purplish, geometric writing. The major did not know what it was, but he did know that it was "nothing from this Earth."

Marcel and his colleagues loaded as much of the material as they could into their vehicles and hurried back to the base, where the debris was examined and then flown to Fort Worth. Marcel escorted the wreckage. He recalled that before he returned to Roswell, he was commanded by General Ramey "not to say anything to anybody"—not even his family. Meantime, the Roswell base commander, Colonel William Blanchard, had reached his own conclusions and had ordered his public information officer to put out the astonishing saucer press release that Ramey demolished with the weather balloon explanation.

Major Marcel's information was the first remotely credible evidence of a UFO crash and wreckage retrieval by U.S. authorities. Until then, ufologists had put little stock in such stories. But Marcel's account, later confirmed under hypnosis and by his family (in disobedience of orders, he had shown them some debris), triggered a fifteen-year quest by UFO investigators for the real story of Roswell.

Stanton Friedman, a nuclear physicist and UFO researcher who was the first to interview Marcel, was among the most avid diggers for information. But Donald R. Schmitt, Director of Special Investigations of the Center for UFO Studies (CUFOS), and Kevin D. Randle, a one-time air force intelligence officer, also mounted a dogged and systematic pursuit, conducting scores of interviews and sifting the information uncovered by colleagues. By 1991, they had put together a mass of what they considered to be corroborative evidence and detail from no fewer than 200 sources, and had published their startling findings and allegations in a book called *UFO Crash at Roswell.*

Schmitt and Randle took testimony from the crews of

the aircraft that continued to fly wreckage to Fort Worth over the course of the next several days. The investigators heard that some of it was transshipped to the Air Technical Intelligence Center at Wright-Patterson Air Base outside of Dayton, Ohio, while other portions went on to Washington. There was a lot of it. On July 9 alone, said one Roswell sergeant, his ground crew labored all day under the gaze of armed guards to load three C-54 transports with crates of material collected from the crash site.

The wreckage turned out to be the least of the developments. Shortly after Major Marcel offered his description of the initial recovery, the UFO investigators got wind of a second crash site, 2.5 miles southeast of the first. Here, witnesses would relate, the desert gave up the remains of what may have been a failed escape capsule, along with the bodies of the alien crew. The beings resembled nothing known on Earth—and the air force had swiftly collected them, too.

ccording to the reports, the air force had been searching frantically for the crew ever since it learned of the saucer crash on July 7. Aerial reconnaissance pinpointed the second site in midafternoon of the next day and a ground party was immediately dispatched. But before the soldiers could get there, first a lone civilian surveyor, then a party of archaeologists had stumbled on the scene. When the soldiers arrived, they cordoned off the area and grimly swore everyone to secrecy. The archaeologists mainly honored their oaths. But the surveyor, one Grady L. Barnett, did not. He told a friend, L. W. Maltais, and the friend after Barnett's death talked to researchers.

What Barnett supposedly saw was a "pretty good sized" dull gray metallic object lying up against the top of a ridge. It seemed to have burst open.

Four bodies were on the ground, all clothed in one-piece suits with no buttons, snaps, or zippers. Barnett described the creatures as being "from another planet." They were only four or five feet tall, with large pear-shaped heads and spindly limbs.

Barnett's report via Maltais might not have been credible by itself. However, in 1989, Mary Ann Gardner, a nurse in a Florida hospital, informed researchers that a terminally ill female patient had confided an amazing tale: The woman said that as a student in the late 1940s, she had been on an archaeology dig in New Mexico and had seen a crashed spaceship and the bodies of the crew. Nurse Gardner had considered it the drug-induced hallucinations of a dying patient—until she happened to see a television show in which the Roswell incident was mentioned.

On February 15, 1990, one of the archaeologists came forward. On condition of anonymity, he averred that upon driving over a rise in New Mexico in July 1947, he had happened upon what looked like a crashed airplane without wings. It was a bulky fuselage, he said, without dome, portholes, or evident hatch; the archaeologist saw no markings. Strewn around the fuselage were three bodies. The one closest to him was small with a big head and eyes; he could see a mouth but could not remember a nose. The body was garbed in a silvery suit and had one arm bent around as if it was broken. The archaeologist said that he and his team were not there long before soldiers arrived.

Issued by Grenada in 1977, these stamps reflect then Prime Minister Sir Eric Gairy's obsession with UFOs. Before he was deposed in 1979, Sir Eric, whose likeness appears on the $2 stamp, thrice proposed to the United Nations—each time unsuccessfully—that it form a committee to study the phenomenon and promote the exchange of UFO information among nations. During his October 7, 1977, address to the General Assembly, Sir Eric claimed that he had himself seen a UFO.

There was more to come. Sergeant Melvin E. Brown was among the soldiers brought in to collect the bodies and clean up the area. On his deathbed years later, he told his family of the experience and described the aliens as smallish, with yellow-orange skin that appeared leathery and beaded. The researchers interviewed his daughter, Beverly Bean, and learned also of a Roswell mortician named Glenn Dennis, who on July 8 had received a number of calls from the base hospital requesting information on body preservation techniques. Curious, the mortician drove out to the hospital. When a nurse found him walking down a corridor, she cried, "My God, you're going to get yourself killed." Immediately thereafter, two military policemen rushed up and carried Dennis outside.

The following day, Dennis told UFO researchers Schmitt and Randle, the nurse explained that three nonhuman bodies had arrived at the hospital and that everyone was in a high state of agitation. She, too, described them as small and delicate, with large heads and eyes. They reminded the nurse of mummies, and something else struck her: The hands had only four fingers, with the two middle digits longer than the others and no opposing thumbs. The nurse said that the bodies had been frozen after examination, then sealed in rubberized "mortuary bags" and packed into a wooden crate for shipment elsewhere. She thought the destination was Wright-Patterson but was not sure.

The UFO investigators were able to find another witness who described how the crate—five feet high, four feet wide, and fifteen feet long—had been loaded into the bomb bay of a B-29 and flown to Fort Worth, Texas. The crew, under the command of Lieutenant Joe C. Shackelford, was ordered to stay at an altitude of 8,000 feet because the bomb bay was unpressurized, and riding with the crate was an armed escort of six military policemen.

The trail petered out at Fort Worth. The researchers found nothing to prove that the bodies had actually gone on to Wright-Patterson. But over the years, Schmitt and Randle uncovered numerous stories of alien remains at the Ohio base. A woman named Helen Wachter recalled that she had been visiting a friend in Dayton in the summer of 1947, when the friend's husband, a guard at the air base, returned home one afternoon in a state of great excitement. Something top secret had taken place—the bodies of four aliens had arrived. He had been on a security detail posted when the plane came in. Later, a contract electrician said he had been asked to check the refrigeration connections in a cold-storage room containing a number of large cases. During the job, he had lifted the lid of one of the cases, and there, on a marble slab, lay a body that was not human.

Others backed up that testimony. One source insisting on anonymity told the CUFOS researchers that he had attended a high-level conference at Wright-Patterson and had seen an alien body in deep-freeze preservation in an underground vault. Just before she died ("Uncle Sam can't do anything to me once I'm in my grave"), a woman named Norma Gardner contacted researcher Leonard Stringfield. She related that as a security-cleared secretary at Wright-Patterson in the late 1940s, she had been assigned the job of logging in all UFO-related material at the base, wherever it came from. In all, she said, she dealt with around 1,000 items, and what she remembered of those from Roswell, including the pathology reports, jibed closely with the testimony of other witnesses.

Nothing in the chain of evidence compiled by CUFOS's Schmitt-Randle team and various colleagues was more startling than the information credited to Dr. Jesse Johnson, identified as the Roswell base pathologist who first examined the bodies, and other

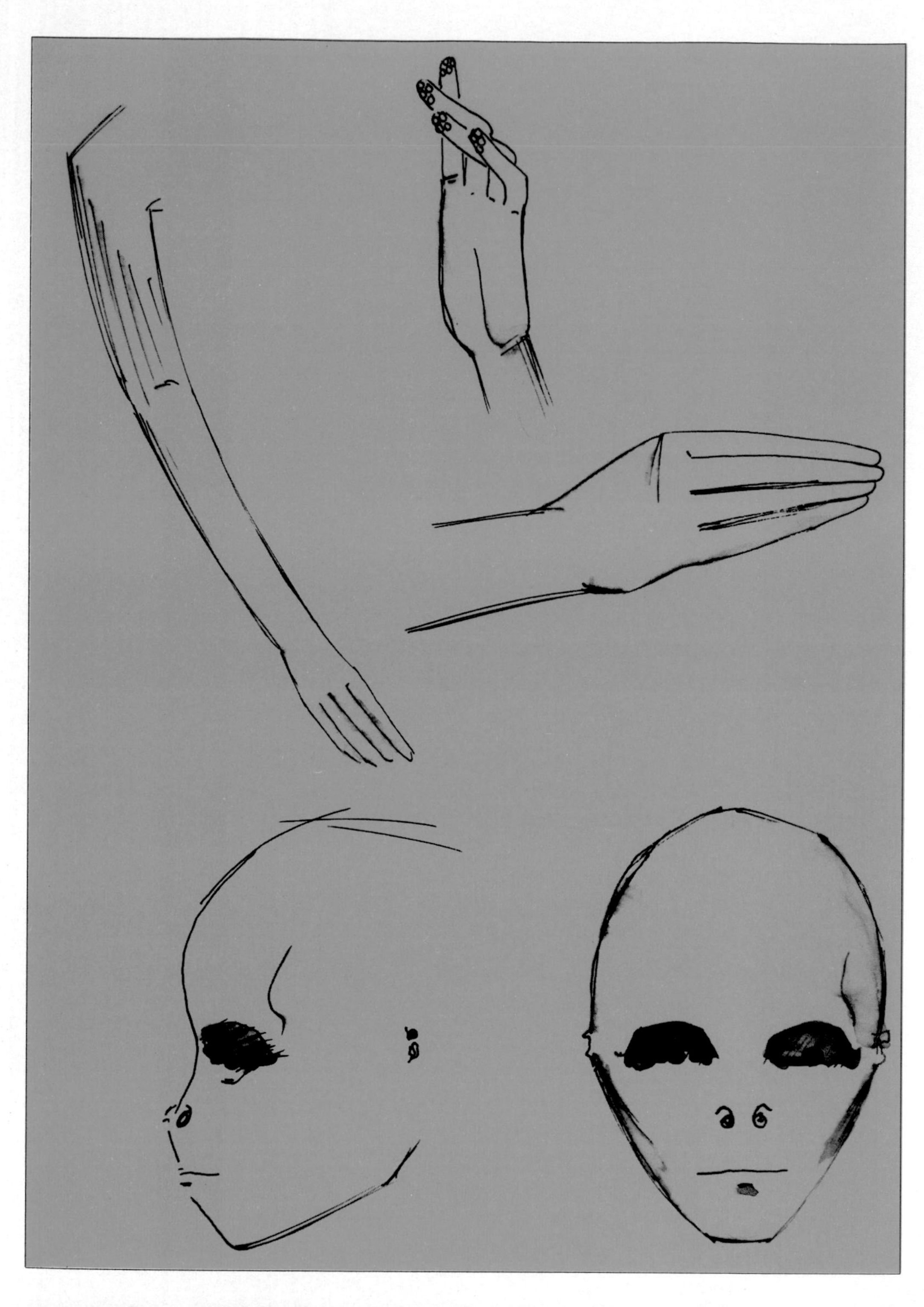

A nurse who said she had assisted in autopsies on three alien beings at Roswell Army Air Field in 1947 reportedly made sketches that became the basis for these illustrations by Roswell mortician Glenn Dennis. The nurse said the aliens' faces were slightly concave and their forearms were longer than humans'. She said their bodies, which had been exposed for days, reminded her of mummies she had seen in pictures.

sources who provided descriptions of the aliens. The bodies were said to be amazingly light for their three-and-a-half- to four-and-a-half-foot length, weighing only about forty pounds. The large almond-shaped eyes had no discernible pupil, nor any real eyelids, just slight folds. Instead of ears, there were small openings in the large head; the nose was indistinct, only a small protuberance, and the mouth was nothing but a slit without lips; it actually seemed to be a wrinklelike fold, scarcely two inches deep and incapable of speech or the intake of food. The arms reached almost to the knees, and the four-fingered hands seemed to have small suction pads at the tips of the digits. By contrast, the legs were very short and birdlike. The skin overall was a pinkish gray, tough and leathery with what under magnification looked like a meshlike structure. A colorless liquid was present in the bodies, but no red blood cells were found, and there was no evidence of digestive or reproductive systems.

No testimony carried more weight—because none came from a more highly placed source—than that of Brigadier General Arthur E. Exon. A World War II pilot with 135 combat missions, Exon was a lieutenant colonel at Wright field in 1947 and commanded the base as a full colonel and then brigadier in 1964 and 1965. CUFOS investigators Schmitt and Randle interviewed him four times between July 1989 and July 1990, twice by telephone and twice in person. The general confirmed the Roswell episode in all of the large and some of the small detail. He said that on arrival at Wright-Patterson, the wreckage had gone straight to the material evaluation laboratories, where it received "everything from chemical analysis, stress tests, compression tests, flexing. The boys who tested it said it was very unusual. They knew they had something new on their hands. A couple of guys thought it might be Russian, but the overall consensus was that the pieces were from space."

Hoping in 1989 to uncover clues to the mysterious flying saucer crash that purportedly occurred at the site in 1947, a team of scientists and researchers from the Center for UFO Studies (CUFOS) explores an isolated stretch of desert some eighty miles northwest of Roswell. Although no crash debris was found on this expedition, CUFOS planned to return to the site for more extensive investigation.

At one point, Exon had flown over the Roswell crash area and had seen the two sites. The bodies, he said, "were all found, apparently, outside the craft itself but were in fairly good condition. In other words, they weren't broken up a lot." When asked if the bodies went to Wright Field, Exon replied, "that's my information." He thought that one of the bodies might have gone to a "mortuary outfit" in Denver, "but the strongest information was that they were brought into Wright-Pat." As for the weather balloon, said Exon, "General Ramey and the people out at Roswell decided to change the story while they got their act together and got the information into the Pentagon and the President." Said the general, flatly: "Roswell was the recovery of a craft from space."

All of this led the researchers to some mind-boggling conclusions: that a vehicle from another world had in fact crashed at Roswell, New Mexico, on or about July 7, 1947; that three, possibly four, beings had perished; that the air force had recovered both the bodies and the debris of the vehicle—and that the government had decided to plow all knowledge of the event deep underground.

The suspicion of official cover-up is never far from the thoughts of ufologists, considering the level of noncooperation exhibited by the authorities. Longtime UFO investigator Richard Hall, who in the 1960s helped direct the prominent civilian National Investigations Committee on Aerial Phenomena (NICAP), suggests a possible scenario that may—if it is accurate—explain something of the why.

Hall says that if the U.S. government has accumulated hard evidence that UFOs are extraterrestrial in origin, and "if strong evidence—or proof—has been obtained that abductions really are taking place, this would amount 'to a secret state of warfare' in which aliens intervene in human affairs at will. Without effective countermeasures or ability to control the situation, political and military leaders would not be likely to reveal the truth." Therefore, says Hall, the U.S.—and presumably other governments—would "practice near-total secrecy while trying to figure out what to do."

tories suggesting government involvement and cover-up do occasionally come to notice. In the early days of the U.S. manned space program, rumors circulated of encounters between astronauts and UFOs. In June 1965, Gemini pilots Edward White and James McDivitt radioed that they had observed an unknown object crossing their orbit. When the report got out, NASA at first said that it was a winged Pegasus satellite—and when that proved incorrect (the Pegasus was more than a thousand miles from the Gemini orbit), the agency simply dropped the subject altogether. Later, in April 1969, Garry C. Henderson, a senior space research scientist working for General Dynamics, was quoted in the press as saying that American astronauts had not only sighted UFOs but had taken photographs—which had been locked away and classified top secret.

At one point in the 1960s, no less a personage than United States Senator Barry Goldwater, wearing a second hat as a reserve air force brigadier general, was rebuffed in an attempt to check out a UFO rumor. Goldwater, as he later told the *New Yorker* magazine, had heard of a secret "Blue Room" at Wright-Patterson where UFO material was

UFO Story That Refuses to Die

On the evening of December 9, 1965, a mysterious object streaked over Lake Erie and reportedly descended to earth in Pennsylvania. Like the alleged UFO crash near Roswell, New Mexico, eighteen years earlier, this incident has never been explained to the satisfaction of some UFO investigators, who continue to explore the case—and to raise allegations of a military cover-up.

Described as an orange fireball, the object was visible to residents of seven states and Canada before it allegedly came down softly in the woods near the village of Kecksburg, about twenty miles south of Pittsburgh. A number of people reportedly rushed to the site, and witnesses who claim to have seen the object say that it was doorless, seamless, acorn-shaped, and ringed with hieroglyphic-like characters. Military and state police officials soon showed up, however, and declared the area off-limits to civilians. The next morning a military truck supposedly hauled away a large dome-shaped object covered by a tarp, bound for Ohio's Wright-Patterson Air Force Base.

Military officials say now—as they did then—that the fireball was a meteor. Pennsylvania UFO researcher Stan Gordon is among many ufologists who believe that the object may have been an extraterrestrial vehicle, which he says would account for the government's brusqueness regarding the affair. But the matter is wreathed in dispute; many residents of Kecksburg maintain that nothing out of the ordinary occurred at all that night. Until witnesses come forward with conclusive evidence—or the U.S. government releases its information, if any—the Kecksburg incident will remain a subject of speculation and controversy.

Army, Police Seal Off Woods In UFO Probe

PITTSBURGH (UPI)—U. S. Army officials and the Penn- ...ia State Police last night ...an area in south- ...nnsylvania explain- ... "is an unidentified ...ect in the woods." ...esman for a team of ...perts from the Army's ...ar Squadron here said, ...n't know what we have

...r sealing off a wooded, ...d area at Kecksburg, Pa., ... 20 miles south of here, ...rs said Army Engineers ... being called to the scene. ...he object was found after a ...h of orange fire streaking ...oss the sky was reported by ...rplane pilots a... ...idents in ...ven states. ... Pentagon s... ...he flash co... meteorite. State P...

'Flaming Ball' Crashes ...outh of Pittsburgh, ...ets Fires in 3 States

Special to The Inquirer

...TTSBURGH, Dec. 9.—A brillia... ...een streaking across sev... ...of fire which ...ay night crashed ... ...d Canada on ...outh of here.

6 States, Canada Report 'Fireball', Sky Rains Debris

PITTSBURG — A flash of orange fire or a "fireball" was sighted in the sky Thursday by airplane pilots and residents in six states and Canada.

Falling debris was reported in Ohio, Pennsylvania and Michigan.

'Fireball' Lands NearPittsburgh; Seen in 7 States

Special to The Inquirer

PITSBURGH, Dec. 9.—A brilliant ball of fire which was seen streaking across seven states and Canada Thursday night crashed into woods 20 miles south of here. Flaming objects falling from it touched off fires in Pennsylvania and Ohio.

The Army and State Police sealed off the wooded area with the explanation:

"There is an unidentified flying object in the woods."

The fireball was seen by airplane pilots and residents of Canada, Michigan, Illinois, Indiana, Ohio, Virginia New York and Pennsylvania.

Early newspaper reports of the Kecksburg affair make no mention of the acorn-shaped object later referred to by some witnesses, but several articles quote army and state police officials as referring to an "unidentified flying object"—a position abandoned soon thereafter. An Ohio man later sketched a domed object, brightly lit and fenced off from approach (left), which he claims he saw while delivering bricks to a hangar at Wright-Patterson Air Force Base a number of days after the alleged UFO descent.

stored. But when the senator asked his friend air force chief of staff General Curtis LeMay, he caught "Holy Hell," as Goldwater put it. LeMay told him that he lacked clearance and that the subject was best forgotten. However, in his autobiography LeMay himself made open reference to UFOs. "There is no question about it," wrote the cigar-chomping general. "There were things which we could not tie in with any natural phenomena known to our investigators."

Yet by and large, those wishing to make a case for official obfuscation and subterfuge have been frustrated in their attempts to find strong evidence. Aside from Roswell, possibly, there is nothing that can be regarded as a "smoking gun." Much of what has circulated might be described as disinformation—stories designed to be proven false or to confuse or to be so far out as to provoke ridicule. Some ufologists have accused the U.S. Air Force Office of Special Investigations of planting spies in the UFO community and concocting stories to make the whole thing seem ludicrous. Tales proliferate that nine races of aliens are visiting Earth. Some stories allege a mad humanoid passion for Tibetan music and strawberry ice cream. Spookier tales whisper that aliens are extracting human enzymes to rejuvenate their own aged and faltering minds and bodies.

In November 1991, a magazine called *UFO Universe* printed an interview with a purported "three-star general." Under the headline "Blowing the Whistle on the Government's Cover Up," *UFO Universe*'s informant described how "the aliens take about 2,200 children a year from the United States and other countries. . . . The children are used in several ways: Biological, to educate and return, experimentation, disease study. Same as adults." Every president of the U.S. "from 1947 to the present day has lied about this situation," the general confided, and worse, "the U.S. Government and the British have made secret treaty agreements with the aliens in exchange for their technology and so-called 'recon' missions during times of human conflict."

Nor is it just the U.S. and Britain said to be in cahoots with the EBEs, or Extraterrestrial Biological Entities, as they are sometimes called. Other tales recount the establishment of a secret American-Soviet-alien space base on the dark side of the Moon, from which thriving human colonies were planted on Mars as early as the 1960s. If that is a trifle hard to swallow, a slightly more digestible top-secret document made the rounds in 1977 describing how aliens had invaded the ICBM base at Ellsworth, South Dakota, and had stolen nuclear missile components while shooting and injuring a guard. The sensationalist *National Enquirer* thought it had a live one there—until its reporters discerned no fewer than twenty errors and other discrepancies in the so-called document. But sensational publications have nonetheless done their witless bit to disqualify UFOs. Consider the London tabloid *Sport,* which regaled readers with the lurid story of how a young boy was turned into an olive by alien ray guns and then consumed in a martini quaffed by the cop investigating the lad's disappearance.

Yet if all the attempts to ignore UFOs or hoot them out of existence have frustrated and pained serious ufologists, the investigators can find encouragement in a new attitude slowly taking hold among a few members of the scientific community. Most scientists have long accepted that since the universe is so gigantic and its stars so numerous—perhaps a billion trillion in the observable universe—it is almost inevitable that other intelligent life besides our own exists out there. But because the distances that separate us from other stars are themselves staggeringly great, scientists on the whole dismiss the possibility that any extraterrestrial beings have visited us.

Nonetheless, some have shown an inclination to listen to colleagues like the University of Arizona's eminent atmospheric physicist, James E. McDonald. Some years ago, McDonald addressed a group of aerospace scientists on the possibility of reaching Tau Ceti, a relatively near-neighbor star system only 11.9 light-years (about 70 trillion miles) distant. "To be sure, we don't have any red hot ideas about getting out to Tau Ceti," said McDonald, "but the pace and tempo of our own technology ought to give pause to those who would insist that there are no Tau Cetians out there who can do that which we still regard as impossible."

A Theory of Alien Beginnings

Some people say extraterrestrial contacts are nothing new. In fact, according to American author Zecharia Sitchin, beings from a faraway planet have influenced human development from time immemorial. Sitchin finds evidence for this astonishing assertion in the epic tales of the Sumerians, founders 5,500 years ago of the world's first high civilization, in what is today the Middle East.

Other scholars consider the Sumerian sagas pure myth. If they have heard of Sitchin's interpretation, which takes the writings literally and then expands on them, they disregard it. Yet this biblical scholar and linguist is one of only about 200 people in the world able to decipher and read the early Sumerian language.

Sitchin has written several books setting forth his theories. In the first, *The 12th Planet,* he says highly advanced aliens brought the human race into existence some 300,000 years ago. Many millennia later, in ancient Sumer (red dot in the satellite photograph of the Earth, above), they passed on the gift of civilization. Sitchin's radical theory of Earth's transcendent alien encounter is explored on the following pages.

Visitors Who Taught the Arts of Civilization

1
2
3
6
5
4
7
8
10
11

The early Sumerians, says Sitchin, called the aliens Anunnaki, or "those who from heaven to Earth came." They revered the celestial visitors as gods, for from them the Sumerians had learned how to live.

The tribes that came before the Sumerians were primitive, their societies having advanced little beyond domesticating animals, growing basic crops, using unsophisticated stone tools, and making utilitarian pottery. When the Sumerians appeared in southern Mesopotamia (present-day Iraq) in the middle of the fourth millennium BC, life changed dramatically. It was as though humankind abruptly grew up.

No one knows for sure where the Sumerians came from, because their language and culture have no traceable antecedents, but they did things that no human beings had done before. They irrigated their farmlands, built magnificent structures, conceived a complex religious faith, practiced law and medicine, and used written language. Scholars call this eruption of civilization "astonishing" and "extraordinary." To Sitchin, however, it makes perfect sense.

He says that the Sumerians were able to step off the plodding track of social evolution and leap far ahead because they had guidance—from the Anunnaki. According to his interpretation of the ancient texts, wisdom was "lowered from Heaven to Earth" by the aliens around 3760 BC. This gift of knowledge, says Sitchin, included everything from medical

The Anunnaki goddess Ishtar (below, left) emerges from a fragment of a wall sculpture that once stood in an Assyrian temple. Sitchin says she wears goggles and earphones festooned with antennas—suggesting she was a member of a technologically advanced society.

According to Sitchin, this nearly 6,000-year-old clay figurine may have been intended to represent a robot emissary of the Anunnaki. He asserts that the ancient texts mention androidlike beings with conical heads who sometimes acted as spokespersons for the gods.

This diagram, based on Sitchin's interpretations, shows the planet Nibiru (12) as it orbits the sun in a vast eccentric ellipse. One of its orbits, according to Sitchin, requires 3,600 Earth-years to complete. While today's astronomers acknowledge nine planets in the solar system, Sitchin says the Sumerians brought the tally to twelve by including the Sun, Earth's moon, and Nibiru. Though modern telescopes have not detected a planet beyond Pluto, many scientists believe that one exists: Both Uranus and Neptune are known to stray from their predicted courses, possibly because of the gravitational attraction of a large but elusive hypothetical planet dubbed by astronomers Planet X.

An Assyrian king reveres the emblems of the chief Sumerian gods in this ancient stone carving (below). The deities numbered twelve—one to rule each planet.

prescriptions to "Earth sciences" and "calculations with numbers." At the same time, the gods, as Earthlings considered them, conferred the tradition of monarchy upon the Sumerians and taught them to create an organized society based upon the concept of universal justice.

The Anunnaki also disclosed the secrets of astronomy. Sitchin says that the ancients actually understood the solar system far better than many later generations of astronomers. For instance, they knew that the Earth was spherical rather than flat and that the Sun, not the Earth, was the center of the solar system, truths that eluded Western astronomers until the Renaissance.

Furthermore, Sitchin says, the Sumerians were aware of all the planets in the solar system that are known today, even those most recently discovered and visible only by means of powerful modern telescopes. They also wrote of a planet that today's astronomers have yet to sight. It was Nibiru, distant homeland of the Anunnaki.

Sumer flourished for about 1,500 years, weathering even a century and a half of political subjugation by its neighbors to the north, the Akkadians. But by 2000 BC, fierce invaders called Amorites and Elamites had snuffed out the remnants of its existence as a discrete nation. Sumerian cultural, political, and social accomplishments lived on, however, to enrich successor civilizations such as Babylonia and Assyria, among whose works much of Sumer's heritage was preserved for the ages. Indeed, Sumerian tales inspired the whole of Western mythology and, some say, even the early chapters of the Bible's Book of Genesis.

undetermined;
Then it was that gods were formed in their midst.

So begins the 4,000-year-old Babylonian *Epic of Creation.* Sitchin writes that these lines, inscribed on clay tablets by an ancient devotee of Sumerian lore, "seat us in front row center, and boldly and dramatically raise the curtain on the most majestic show ever: the Creation of our solar system."

In a controversial interpretation of the epic, Sitchin professes to reveal the truth about how the solar system was born. To him, the deities named in the saga are planets, and the story of their struggles can be taken as a representation of a credible theory of cosmology.

In the beginning, he says, there were three heavenly bodies in our solar system: Apsu, Mummu, and Tiamat (opposite, top, 1, 2, and 5, respectively). Apsu was the Primeval Father, the Sun, and Mummu his trusted acolyte, the planet Mercury. Tiamat was the Mother Goddess, and she and Apsu were said to spawn the planets we know as Venus, Mars, Jupiter, and Saturn (3, 4, 7, and 8, respectively). The texts do not say how. Jupiter and Saturn

system concedes little in accuracy to the modern diagram above it, proving, he says, the ancients' understanding of astronomy. Both diagrams arrange the planets counterclockwise in order of distance from the Sun, beginning with Mercury. The only significant differences—here indicated by dashed lines—are the position of Pluto and the inclusion of Nibiru, the twelfth planet.

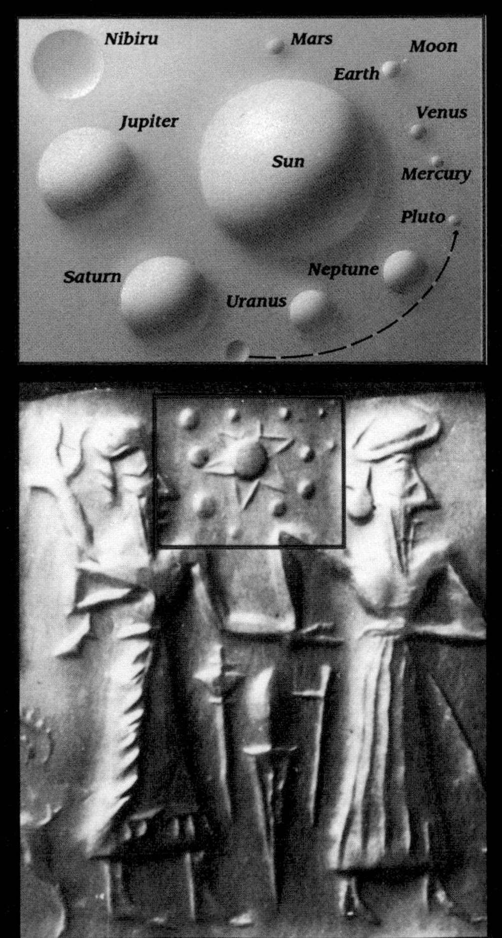

est and most favored satellite, Kingu (6). The young gods wanted revenge, but they feared themselves too weak.

Saturn perceived that Nibiru was sufficiently strong to smite the hated Tiamat and pleaded with him to do so. Nibiru agreed, on the condition that the other planet-gods award him supremacy over them. They consented and adjusted their own orbits so as to place him on a collision course with Tiamat.

Armed with fire, lightning, and a small army of "winds"—satellites, according to Sitchin—Nibiru drew toward the Mother Planet. When he came within reach, he "split her into two parts" like "a mussel" (diagrams opposite). Tiamat was no more.

All her former satellites were condemned to wander the far reaches of heaven—except for the once mighty Kingu, whom Nibiru rendered "a mass of lifeless clay." The detritus from Tiamat's destruction, including a large piece that became the Earth, was soon corralled into assigned places in the solar system. Finally, asserting his dominion over the other planets, Nibiru dispatched them to their current orbits. He then "crossed the heavens and surveyed" his handiwork. The solar system was complete.

1 4 5 6 2 7 10 9 8 11 12

New path of Nibiru

Original path of Nibiru

Pulled from its preexisting course (broken line) by the force of Neptune (11), Nibiru (12) enters the solar system from "the heart of the Deep" and makes an orbital beeline for Tiamat (5), according to Sitchin's version of the Epic of Creation. Called the Planet of the Crossing because its path intersects the orbital planes of some of the other planets, Nibiru moves in a clockwise course while the rest travel counterclockwise.

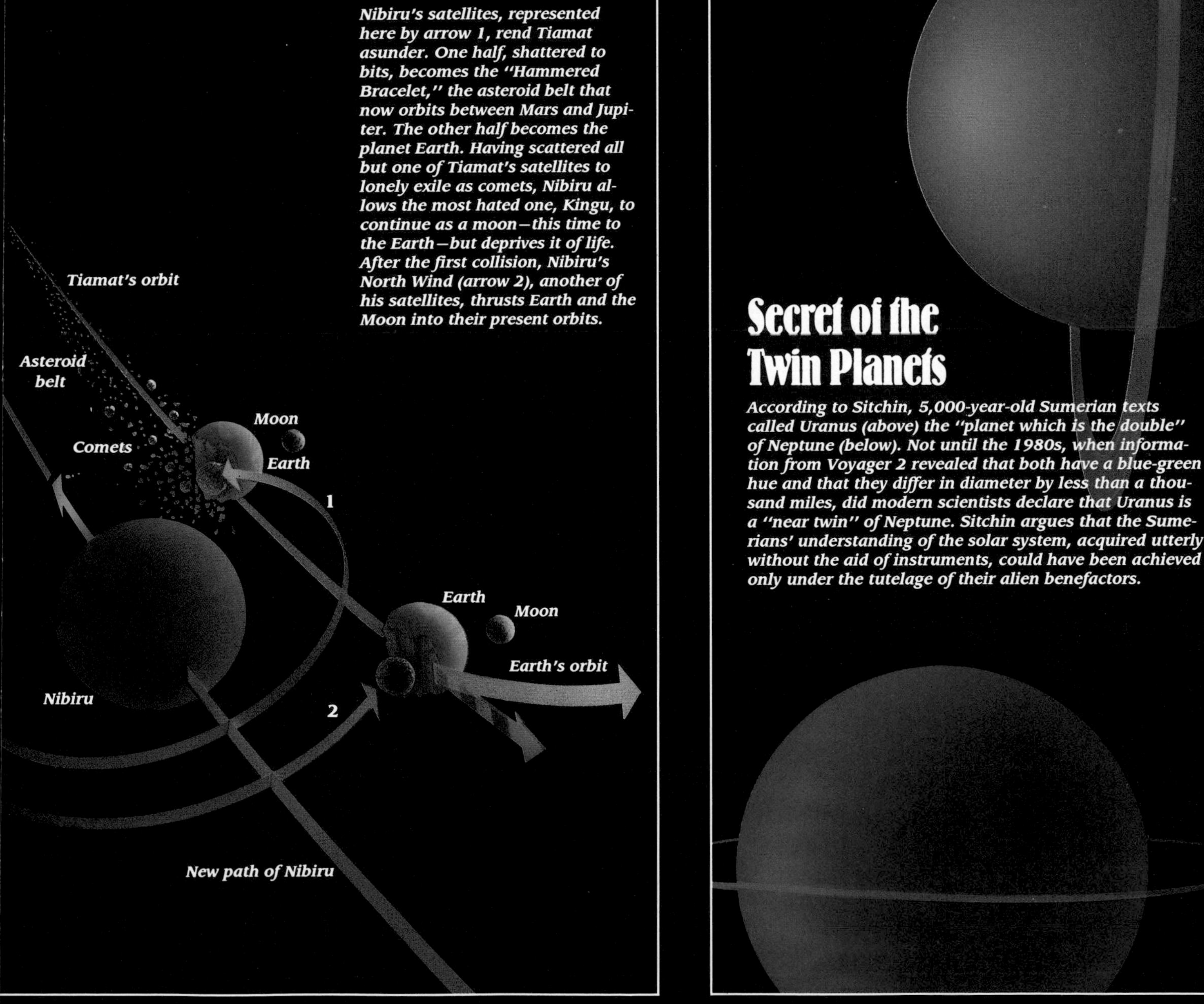

Nibiru's satellites, represented here by arrow 1, rend Tiamat asunder. One half, shattered to bits, becomes the "Hammered Bracelet," the asteroid belt that now orbits between Mars and Jupiter. The other half becomes the planet Earth. Having scattered all but one of Tiamat's satellites to lonely exile as comets, Nibiru allows the most hated one, Kingu, to continue as a moon—this time to the Earth—but deprives it of life. After the first collision, Nibiru's North Wind (arrow 2), another of his satellites, thrusts Earth and the Moon into their present orbits.

Secret of the Twin Planets

According to Sitchin, 5,000-year-old Sumerian texts called Uranus (above) the "planet which is the double" of Neptune (below). Not until the 1980s, when information from Voyager 2 revealed that both have a blue-green hue and that they differ in diameter by less than a thousand miles, did modern scientists declare that Uranus is a "near twin" of Neptune. Sitchin argues that the Sumerians' understanding of the solar system, acquired utterly without the aid of instruments, could have been achieved only under the tutelage of their alien benefactors.

Colonists from Outer Space

Nibiru, according to Sitchin's interpretation of the texts, was fertile with "the seed of life" when it created Earth by smashing Tiamat to pieces. Some of these vital ingredients cleaved to the Earth, producing eons later a rich and varied array of flora and fauna. On Nibiru itself, over billions of years after that fateful collision, the life-giving material grew into a full-fledged civilization of astoundingly advanced beings—the Anunnaki.

About 450,000 years ago, or 125 Nibiruan years—that is, orbits around the Sun—the Anunnaki discovered that their continued existence was in jeopardy: Nibiru's atmosphere was slowly dissipating. Under pressure to save their world, according to Sitchin's reading, Anunnaki scientists devised a plan. They would create a vast protective shield of gold particles to suspend above their dwindling atmosphere.

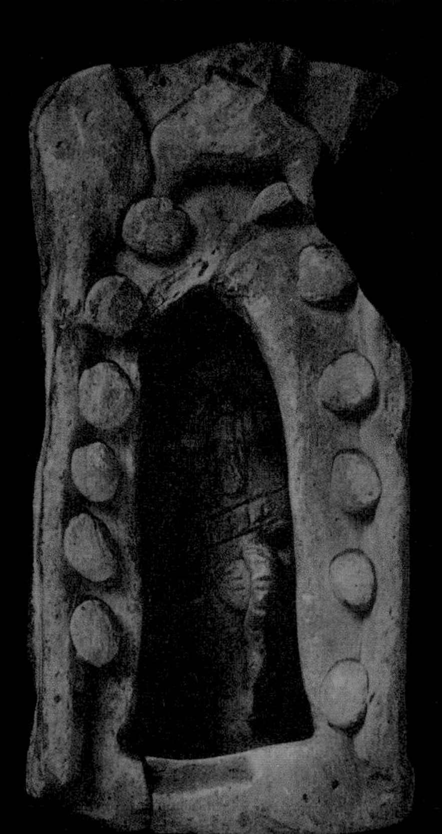

The Anunnaki knew where to go to find gold in great quantities. Probes had revealed that Earth was the only planet in the solar system with rich veins of gold. Wasting no time, the Anunnaki planned a prospecting expedition.

They were forced to wait until their immense orbit brought them to an opportune position to launch a flight to Earth. They reached the planet at a time when Earth's second glacial period had covered a third of it in ice. Undaunted, the extraterrestrial colonists set course for the region that was later called Mesopotamia. It seemed an ideal base: Still temperate despite the Ice Age and plentifully watered by the Tigris and Euphrates rivers, it was also richly endowed with petroleum for fuel. Most important, Mesopotamia was close to the Persian Gulf, from whose waters they planned to extract the gold they had come for.

In the artifact at left, an Anunnaki stands inside what Sitchin terms a rocket ship—because, he says, the Sumerians called it a "sky chamber." Twelve globes surround the chamber, one for each planet. In an illustration (above) taken from a Sumerian cylinder seal, a winged spacecraft hovers between an inbound astronaut, on the left, and an Anunnaki on Earth. Mars is represented by the six-pointed star; Earth, by the seven globes, as it is the seventh planet in from Nibiru. The crescent is Earth's moon.

The first spacecraft splashed down in the Arabian Sea, the northernmost extension of the Indian Ocean. The astronauts made their way to land and trekked north. In those days, the northern end of the Persian Gulf was a swamp, and on that marshy land, the Anunnaki established Earth's first city, Eridu. The ancient Sumerian name literally means "house in faraway built." More settlements followed, in a pattern that would form a landing corridor visible to incoming astronauts far out in space. The Anunnaki kept shuttlecraft orbiting Earth as intermediaries between ships from Nibiru and the settlements on Earth. They also probably had some sort of way station on Mars.

The leader of the colony, who appears in the old tales as a god named Enki, maintained his seat of power at Eridu, but his reign over Earth was short-lived. He failed to obtain sufficient gold from the waters of the gulf, and his father, Anu, replaced him with Enlil, his half brother. Sitchin attributes to the Anunnaki life spans appropriate to their planet's long orbital years. Thus, 28,800 years (8 Nibiruan years) after the first landing, Enki was forced to cede his power.

By this time, a warming climate melted the great icecaps, raising the level of the Earth's seas and flooding the swamplands of the Anunnaki's first settlements. The colonists moved farther inland, to the center of Mesopotamia, where Enlil inhabited a settlement called Larsa while his new capital, Nippur, was being built.

Some 21,600 Earth-years in the making—a mere 6 years to the beings from Nibiru—Nippur became, in Sitchin's words, "a sophisticated command post from which the Anunnaki on Earth could coordinate journeys to and from their home planet, guide in landing shuttlecraft, and perfect their takeoffs and dockings with the spaceship orbiting Earth."

Given the lack of success he had in harvesting gold from the waters of the Persian Gulf, Enlil thought to search for the precious metal on land. This decision was destined to change the course of life on Earth for eons to come.

Sitchin believes that Mesopotamian ziggurats such as this one served as observatories and as landing markers for Anunnaki spacecraft. The stepped pyramids were equipped, says Sitchin, with ringlike antennas for communication with orbiting shuttles.

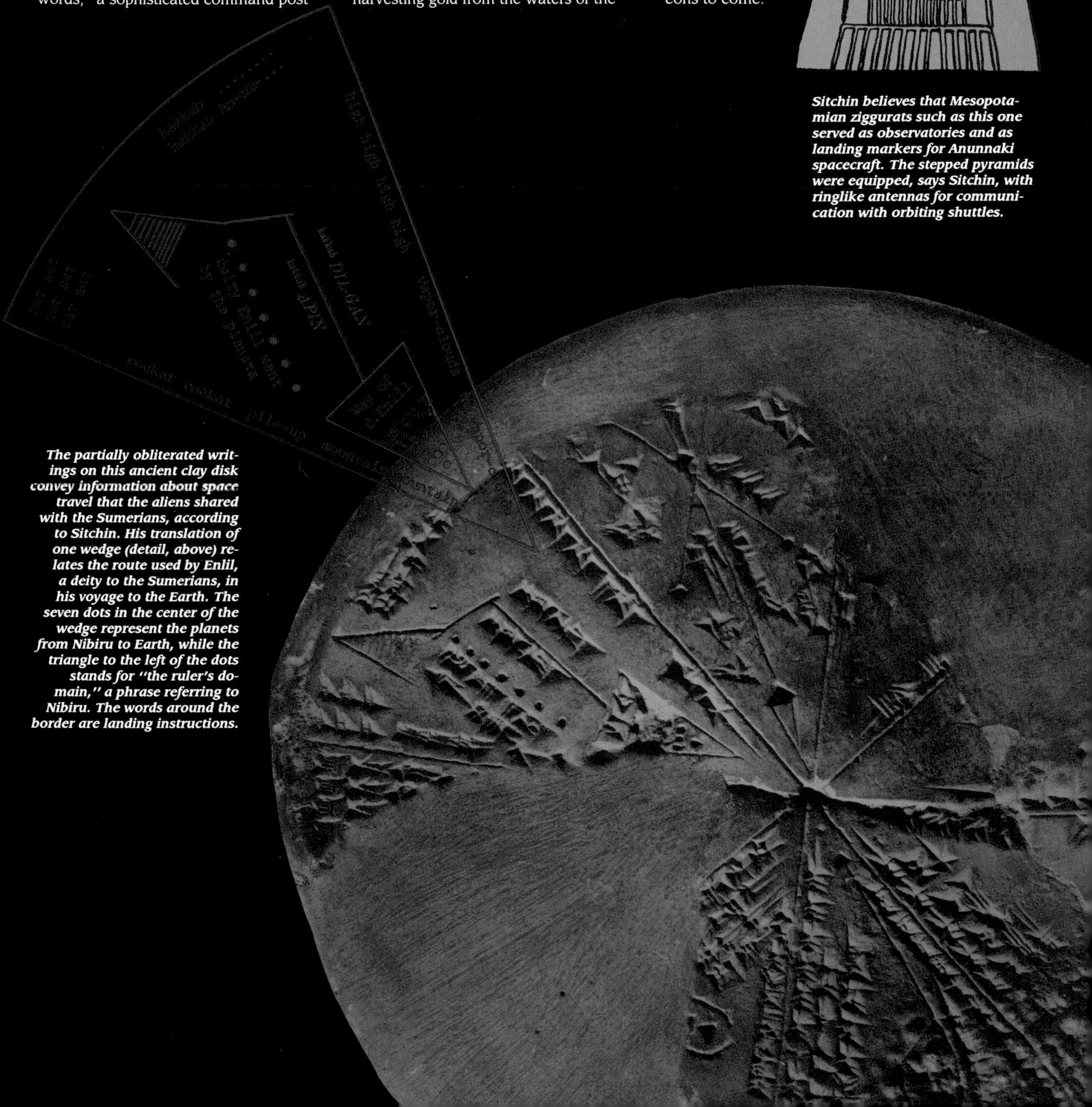

The partially obliterated writings on this ancient clay disk convey information about space travel that the aliens shared with the Sumerians, according to Sitchin. His translation of one wedge (detail, above) relates the route used by Enlil, a deity to the Sumerians, in his voyage to the Earth. The seven dots in the center of the wedge represent the planets from Nibiru to Earth, while the triangle to the left of the dots stands for "the ruler's domain," a phrase referring to Nibiru. The words around the border are landing instructions.

Making Humans in the Laboratory

Enlil's quest for gold ended in a lush and grassy land far from Mesopotamia. Sitchin believes this place to have been southern Africa. He says the Anunnaki sailed to present-day Mozambique. In mines there, the astronauts grew tired of the back-breaking toil and uncomfortable working conditions. The situation became so intolerable to them that when Enlil visited the mines, they staged an impassioned mutiny, described by the ancient texts as a rebellion of deities.

"Every single one of us gods has war declared!" they shouted. "Excessive toil has killed us, Our work was heavy, the distress much." Enlil was not moved, but his father, Anu, sided with the mutineers, as did Enlil's rival half brother, Enki. Enki proposed that he and Ninharsag, the goddess in charge of medicine, create a *lulu,* a primitive worker, to relieve the gods of their toil.

They combined the genes of birds, cattle, lions, and other Earth animals with those of a being that seemed a cut above the rest: the ape-man, or hominid. The results were disappointing *(top right).* Then they created a satisfactory lulu—the first human—by mixing the ape-man's genetic material with that of the Anunnaki. Ninharsag, thereafter named Ninti, or "lady who gives life," held up the prototype hybrid babe *(lower right)* and cried out, "My hands have made it!"

The lulu resembled the Anunnaki. Unlike its hairy ape-man antecedent, the hybrid had, according to one ancient text, "a skin as the skin of a god." The first humans were created infertile and were turned out en masse by the Anunnaki, with many birth goddesses simultaneously incubating fetuses.

Sitchin writes that the idea of alien genetic engineering solves a problem in the theory of evolution—the missing link. Anthropologists are able to trace the development of apes into humanlike apes and humanlike apes into apelike humans. But when the scientists reach the immediate antecedents of *Homo sapiens,* they run into trouble. In Sitchin's words, "*Homo sapiens* represents such an extreme departure from the slow evolutionary process that many of our features, such as the ability to speak, are totally unrelated to the earlier primates."

The answer, he claims, is extraterrestrial intervention. If the Anunnaki could set up sophisticated establishments on a planet not their own, he reasons, they might be capable of genetically altering beings, producing a great leap in the process of human evolution.

Sitchin says the Book of Genesis is simply a revision of the older Sumerian stories. According to the Epic of Creation—the Babylonian version of a Sumerian text—the humans were imported to Mesopotamia, the land many scholars associate with the biblical Garden of Eden. There, Enlil and Enki, bitter enemies since Enlil supplanted his brother as Anunnaki chief on Earth, struggled against each other for supremacy.

Enki, part creator and longtime ally of the humans, decided to thwart his brother's ironfisted control over them. Like the serpent in the biblical story, he encouraged the humans to sample a forbidden fruit: He gave them the ability to procreate. Much as Adam and Eve became aware of their nakedness, the humans discovered in the gift of reproduction the power to rule their own destiny. Enlil, enraged and fearful that the humans might also learn the secret of immortality, cast them out of Eden to fend for themselves.

Banished, the humans continued to procreate, eventually even intermarrying with the gods, according to Sitchin's translation of the original Hebrew Book of Genesis. "And it came to pass, when the Earthlings began to increase in number . . . and daughters were born unto them, that the sons of the deities saw the daughters of the Earthlings . . . and they took unto themselves wives of whichever they chose." Vehemently opposed to such intermingling, Enlil plotted to purge humankind from the planet.

The Anunnaki knew that a natural disaster was on the way. Nibiru would soon pass near Earth on its long orbit around the Sun, and they suspected that its gravitational pull would destabilize Antarctica's ice sheets and cause them to slip into the oceans, flooding the planet and drowning all life. As the fearsome time approached, the Anunnaki, under Enlil's direction, took refuge in space shuttles that orbited Earth, without warning the Earthlings.

Little did they know, however, that Enki, champion of humankind, had betrayed their ill intentions by telling a wise man, called Utnapishtim in an Akkadian text, of the coming disaster. In response, the human—whom Sitchin identifies as the biblical Noah—built a large vessel and stocked it with plants and animals of all varieties. In this manner, humankind survived the Deluge.

When the waters receded, the gods returned to Earth and were surprised to find Noah on Mount Ararat. Enlil was outraged at first, but then welcomed the survival of the humans, realizing how important they would be in the immense task of reconstruction that lay ahead. The Anunnaki promised never to harm their progeny again. And humankind, no longer in bondage, would henceforth become a partner on Earth. The gods began to grant the Earthlings knowledge and rudimentary social organization, ultimately conferring upon them, in Sumer, the great gift and responsibility of an advanced civilization of their own.

These Sumerian illustrations reveal the various stages in the Anunnaki's creation of humankind. Anunnaki god-laborers (far left) toil in the mines of southern Africa before the humans were created to relieve them. A gallery of failed hybrids (center), half-animal and half-hominid, shows what Ninharsag and Enki invented and discarded before they arrived at humankind. The goddess Ninti (near left), with the tree of life to her right, brings forth the first human. Beakers and test tubes offer evidence of the genetic laboratory that produced the babe.

CHAPTER 3

The Quest for Explanations

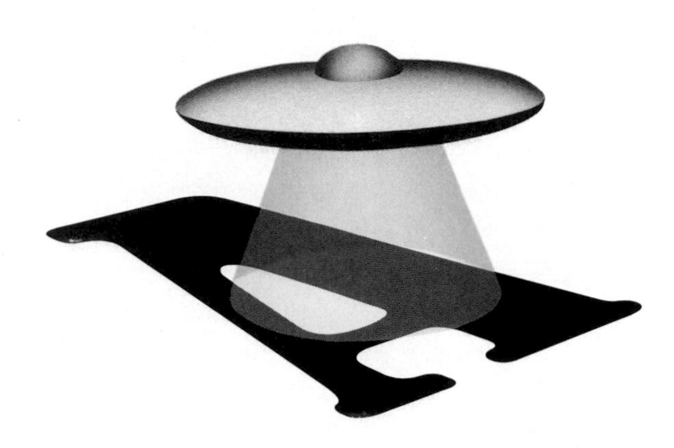

round the world, small towns have a reputation for gossip, and in 1975 the Swiss village of Hinwil was no exception. Speculation was rife that year about the puzzling activities of an unemployed watchman named Eduard Meier, usually known as Billy. A soft-spoken man with a long beard and piercing hazel-green eyes, Meier was already peculiar enough by Hinwil standards. After decades working at various jobs throughout Europe and southern Asia, he had come back to Switzerland a few years earlier with a young Greek wife, three children, and only one arm, having lost his left arm in a bus accident in Turkey. Now this unusual man was behaving in a still more remarkable way.

Ever since January, Meier had been leaving his house at all hours of the day and night, puttering off into the Swiss winter on his moped and staying in the nearby woods for hours. For more than a year, his baffled neighbors could only watch and wonder, as Meier skillfully fended off any questions about his jaunts. Then the July 8, 1976, issue of the popular German magazine *Quick* appeared, and the truth—or what Meier said was the truth—was out. According to the magazine, Meier said that he was going to the forest to meet with alien beings from the Pleiades, conveyed to Earth in vehicles he called "beamships." Moreover, Meier told the magazine's reporters, he had photographs, dozens of photographs, to prove it.

His usual contact, Meier said, was a space creature named Semjase, an amber-haired woman of luminous beauty. According to Meier, Semjase was an emissary from the planet Erra, located in the star cluster of the Pleiades. The people of the Pleiades, she told him, lived to be 1,000 years old. Semjase herself admitted to being just 330 years old. Despite her relative youth, Semjase had an important mission, Meier reported: She was here on Earth to lead humanity toward a path of higher spiritual evolution. "We endeavor to keep order throughout all areas of space," she told Meier, and help was clearly needed. For, as Semjase explained, the inhabitants of Earth had already twice destroyed their planet in ages past, by insane warfare and technology gone wild.

Semjase's revelations were only part of Meier's story, which gradually emerged in subsequent interviews and later became the subject of a 1987 book, *Light Years,* by American journalist Gary Kinder. Meier contended that his alien encounters near Hinwil were just the latest in a series of contacts he had experienced at eleven-year intervals, beginning in 1942, when he was five years old. At that time, Meier had supposedly met a space creature named Sfath, whose voice was "like a gentle and fine laughter." Meier said his father had even seen Sfath's spaceship—but had dismissed it as a new type of Nazi aircraft.

As Meier told the story, Sfath remained in mental contact with him for eleven years, until 1953. After a brief hiatus, Meier then received a second mentor, Asket, from a parallel universe known by the initials DAL. Among other communications, she informed Meier that his ancestors had themselves been extraterrestrials from the constellation Lyra. Asket remained on duty until 1964, eleven years after her first appearance. (Although Asket's existence, like that of Sfath, remains unprovable, Gary Kinder notes that there is evidence that Meier claimed to be in contact with aliens in 1964. In the summer of that year, the New Delhi *Statesman* ran a story about an eccentric Swiss near the Indian village of Mahrauli—where Meier was then working—who said he had photographed alien ships.)

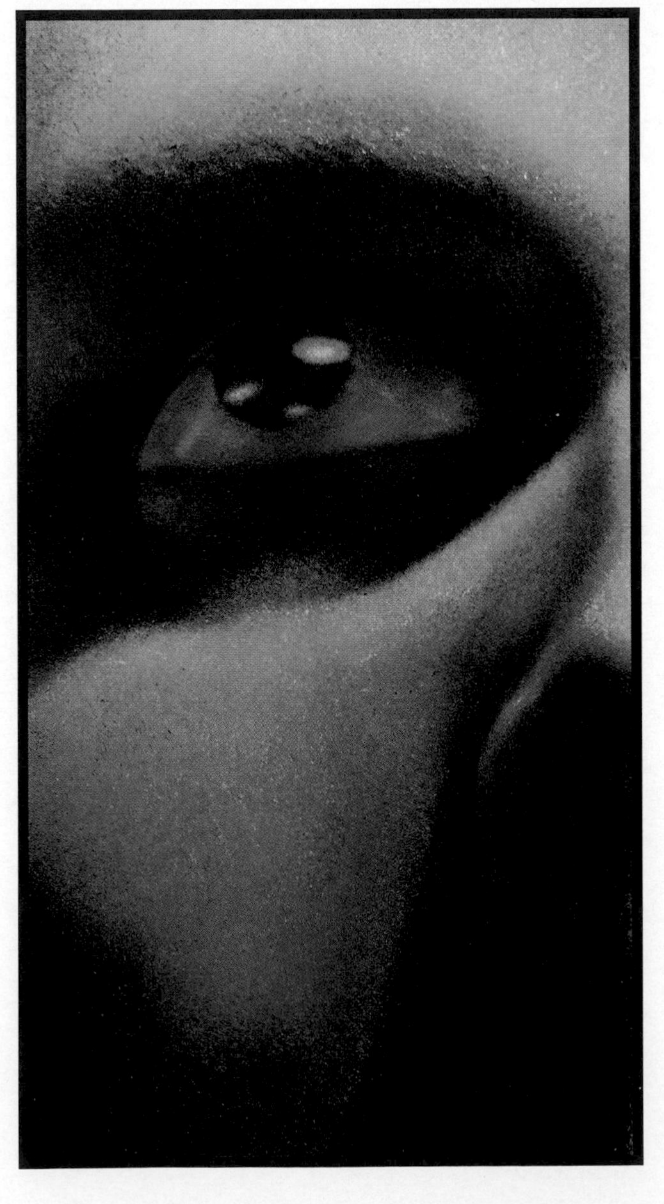

In these earlier contacts, Meier said, the aliens told him he was not yet ready to receive their spiritual instruction. But with the appearance of Semjase in 1975, eleven years after Asket's departure, the time for learning had finally arrived. In fact, according to Meier's later statements, the Pleiadians would continue to meet with and teach him well into the 1990s.

Throughout that extended period of contact, during which Meier and his impoverished family moved from Hinwil to a derelict farmhouse in the nearby town of Schmidruti, Meier produced many additional proofs of his story, including more photographs, several eight-millimeter movies, audiotapes of the beamships' engines, some metallurgic samples supposedly provided by the Pleiadians, and thousands of pages of transcripts of his conversations with Semjase and her companions.

Neighbors and friends who dropped Meier off on the way to his woodland rendezvous also had some odd tales to tell—of Meier reappearing out of nowhere, warm and dry in the midst of pouring rain, or of strange lights that skittered overhead, too fast-moving to be airplanes and too neatly formed for flares or fireworks. A few people heard what they guessed might be the sound of approaching spacecraft, described by one listener as "an eerie and grating noise, like a high-pitched cross between a jet airplane and a chain saw."

As word of Meier's contacts with the Pleiadians continued to spread, a flood of terrestrial visitors descended on Hinwil and Schmidruti: journalists, ufologists, students of psychic revelation, and other seekers who included the American

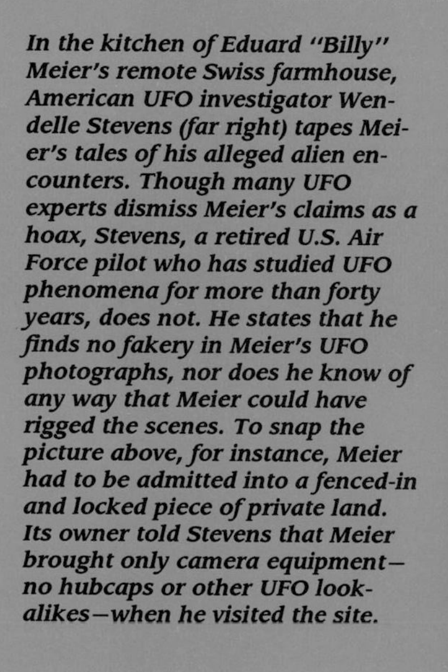

In the kitchen of Eduard "Billy" Meier's remote Swiss farmhouse, American UFO investigator Wendelle Stevens (far right) tapes Meier's tales of his alleged alien encounters. Though many UFO experts dismiss Meier's claims as a hoax, Stevens, a retired U.S. Air Force pilot who has studied UFO phenomena for more than forty years, does not. He states that he finds no fakery in Meier's UFO photographs, nor does he know of any way that Meier could have rigged the scenes. To snap the picture above, for instance, Meier had to be admitted into a fenced-in and locked piece of private land. Its owner told Stevens that Meier brought only camera equipment—no hubcaps or other UFO look-alikes—when he visited the site.

actress and New Age enthusiast Shirley MacLaine and European paranormal investigator Erich von Däniken. One key visitor was a second cousin of Carl Jung named Louise Zinsstag, a prominent European UFO researcher who was then in her seventies. Without determining whether Meier was truly in touch with extraterrestrials, Zinsstag declared that Meier was probably a witch and certainly "the most intriguing man I have ever met."

Although Meier seemed overwhelmed at times by the never-ending stream of callers, he answered their questions willingly. Sometimes he took visitors to the alleged rendezvous sites, where he pointed out the beamships' circular landing tracks—swirls of crushed grass that reportedly neither withered and died nor grew upright again.

Even as Meier's fame grew, however, so did rumors that he might be faking. The sheer elaboration of his story worked against him, and his vast accumulation of evidence seemed to some investigators more impressive for its quantity than its quality. A few critics questioned why Meier seemed so willing to share his experiences with others, when many, although not all, of those said to have encountered aliens come forward only reluctantly. By the early 1990s, opinion remained divided on the Meier case. Although some UFO researchers were intrigued by his saga, many ufologists believed that the one-armed Swiss villager was a remarkably talented fraud.

Yet few who spent time with Meier could doubt the strength of the man's conviction that he had been chosen to act, as he put it, as "truth offerer" to the world. To Meier's followers—and by the early 1990s he had hundreds, in many countries—the quiet Swiss villager retained an almost mystical status as a primary witness to one of the planet's most astonishing realities.

The debate over what happened to Billy Meier—and what it might mean—encapsulates many of the theories about UFO sightings and alien encounters. The questions to be answered are basic indeed: What are UFOs? Mass hallucinations? Natural phenomena? Or real spaceships? If they are real, where do they come from? Have human beings been abducted? Or have they simply imagined—or invented—such abductions? Like the investigators, journalists, and parapsychologists who flocked to Meier's Swiss home, other researchers whose backgrounds range from engineering to psychology, from astrophysics to the study of folklore, have spent thousands of hours digging into cases of reported alien encounters. Still the answers remain elusive.

Some investigators—confirmed skeptics—take their cues from such scientists as the late Donald Howard Menzel, a Harvard astronomer who argued in the 1960s that every UFO sighting that is not an outright fraud is attributable to natural phenomena such as a meteor shower or some unusual atmospheric effect. According to this line of reasoning, memories of an alien encounter can be no more than the mental aberration of an overwrought observer.

A more sympathetic psychological interpretation is one first propounded by none other than Carl Jung, Louise Zinsstag's illustrious second cousin. A famed psychiatrist who studied under Sigmund Freud and then went on to found his own system of psychoanalysis, Jung saw tales of alien visitors as springing from deep within the collective soul of humankind. UFO space travelers were an emanation of ancient terrors and desires as updated for twentieth-century minds, he suggested, angels and demons in space-age costume. This equivalency cuts both ways. Either alien visitors are modern myths, or, as Jung himself admitted, flying saucers could be actual, nuts-and-bolts spacecraft that have given rise to the archetype through the ages.

The latter alternative, of course, is the one favored by many enthusiastic ufologists, who contend rather straightforwardly that the UFO phenomenon is exactly what it seems. UFOs, they say, are indeed alien craft from outer space, piloted by beings with considerably more brains and technological savvy than ourselves.

Such researchers are divided on the issue of alien encounters, however. Some ufologists see abduction reports as complex, often absurd fantasies that interfere with legitimate UFO research—or as essentially irrelevant evidence of some completely different phenomenon. England's Hilary Evans, for example, suggests that aerial UFO sightings and alien abductions should be researched separately. Other investigators hold just the opposite view. They believe that reports of close contact with aliens offer the key to understanding the UFO phenomenon—to answering the question, What do the aliens want? "Abductions have cracked open the UFO mystery like a cosmic egg," declares historian David Jacobs. "Inside we see alien life, the creation of bizarre life, and the exploitation of human life," a reference to theories that aliens are conducting genetic experiments on selected human beings.

till others hypothesize that the UFOs are real but are not simple spaceships. Physical laws place formidable barriers in the way of practical interstellar travel, suggesting to some that the so-called aliens might really be visitors from Earth of the far future, reaching the present via a cosmic time warp. Or perhaps UFOs are from an alternate universe, another reality beyond normal space and time. Turning from advanced physics to parapsychology, yet another school of thought holds that UFOs, and other alien presences, may be purely psychic phenomena, as tantalizingly real or unreal as ghosts or guardian angels. Louise Zinsstag may have been hinting at this solution to the Billy Meier case when she called him a witch.

If they can be believed, still another group of people—neither UFO researchers nor inadvertent eyewitnesses—offer their own inside perspective on the phenomenon. For decades, mediumistic channelers have been said to pass on elaborate messages from alien space travelers. And their fantastic communiqués pale beside yet another claim: that hundreds of human beings, perhaps including Billy Meier, are "star people" descended from aliens, here on Earth to educate human beings for a great change to come.

Given such a crowded galaxy of theories, it is hardly surprising that many active ufologists make it a point to keep an open mind. Stressing that he believes the UFO phenomenon is bound to produce "scientifically important discoveries," Michael Swords, a professor of natural sciences at Western Michigan University and editor of the *Journal of UFO Studies,* wrote in 1989 that he thought all theories were premature: "Let the data, not the opinions, talk."

The trouble is, of course, that the data can be frustratingly difficult to pin down. UFO enthusiasts and skeptics alike agree that the body of UFO and abduction reporting is plagued by fraud in at least some cases, although a lack of research funds and adequate staff makes it difficult to be certain when fraud has occurred, how it happened, and why. The Meier story is an excellent case in point. By some assessments, no episode of alien contact has generated a greater volume of recorded evidence—and none is more racked by controversy.

Meier's photographs, for example, are in many ways hard to fault; they remain among the sharpest, most convincing UFO pictures ever taken. Most of them show one or more of the Pleiadian beamships, silvery, disk-shaped objects about twenty-one feet in diameter with rectangular red windows encircling their raised centers. Over the decades, Meier has accumulated several hundred colorslides showing six distinct types of beamships aloft near familiar trees and mountains.

Probably inspired by early explorers' tall tales of their journeys, this fifteenth-century Italian print (above) portrays peoples that supposedly inhabited faraway lands. From left to right are the sometimes headless Musteros of North Africa or Asia; the tiny Pygmy hunters of Africa, and the huge-footed Sciopedi of India. In our own time, though global travel and communication have demystified most of the inhabitants of this planet, humankind continues to speculate about the unknown beings that might dwell on other planets. At right, an early-twentieth-century illustration brings to life English writer H. G. Wells's 1907 description of Martians—which bear a strong resemblance to aliens described by supposed abductees nowadays. Wells wrote that the denizens of the red planet possessed great intelligence, which necessitated their having "big brains" encased in "big shapely skulls."

Reaction to the photographs has been mixed, in part because most of the originals and negatives are missing. In the chaos of the Schmidruti farmhouse and the constant onslaught of curious visitors, Meier says, many of his photographs and other proofs have been either misplaced or stolen. Photography experts do not find copies as convincing, since clever manipulation during the copying process could erase any telltale signs of fraud.

Despite the lack of originals, however, some authorities remain impressed by Meier's work. Neil Davis, a physicist whose photo-optical company worked for the U.S. Navy, saw none of the usual evidence of hoaxing, according to Gary Kinder's account of the Meier case. Images of the supposed beamships and their surroundings are equally sharp, are illuminated identically, and appear in the same density, Davis said, suggesting Meier had not simply superimposed two pictures in a double exposure. "From a photography standpoint, you couldn't see anything that was fake about them," agreed Bob Post, head of the photo-developing staff at the Jet Propulsion Laboratory. Michael Malin of Arizona State University called Meier's images "by far the best UFO pictures taken," but added skeptically "whether they're authentic or not, that's a totally different matter."

On the grounds of probability, those who credit Meier's account argue that an impoverished one-armed man with no access to a photography studio would be hard-pressed to create faked photographs so resistant to photoanalysis. When Wally Gentleman, a Hollywood special-effects technician, reviewed Meier's films and several of the photographs, he concluded that only a team of about fifteen professionals could have produced images so seemingly authentic. "If somebody is faking them, they have an expert there," he commented. "If the expert knowledge isn't there, this has got to be real."

And yet Meier, despite his disability, is no slouch at darkroom legerdemain. One researcher, for example, found some partly destroyed slides showing a carved wooden model of a beamship. Asked about the slides, Meier claimed that he had only been testing his photographic skills and had managed in the process to prove that the real beamship photos could not be duplicated through fakery.

If the photographs present a mixed set of evidence, the tape recordings of the beamship engine noises receive more positive reviews. Again, the entertainment industry, with its experience in the sheer difficulty of illusion making, is most enthusiastic. Given samples of the tapes, electronics specialist Steve Ambrose, sound engineer to rock singer Stevie Wonder, was at a loss to explain the intricate, seemingly natural, layering of audio textures and frequencies. "If it is a hoax," Ambrose said, "I'd like to meet the guy who did it, because he could probably make a lot of money in special effects."

As for the metallurgic samples, all but one proved under analysis to be familiar earthly materials—amethyst, pot metal, silver solder, and so on. The exception, a burnished golden-silver triangle, was more intriguing. Marcel Vogel, an IBM research chemist with a lifelong interest in topics on the fringe of conventional science, examined the specimen at a UFO researcher's request. He found it included discrete samples of many purified substances, including one extremely expensive, rare by-product of atomic energy work.

Unfortunately, Vogel had no sooner completed his examination—which he preserved on videotape—than the precious sample vanished. According to Vogel, he placed the triangle in his lab coat pocket, but the next morning it was not there. The only irrefutable physical proof of Meier's story, if such it was, is apparently gone for good.

Other evidence in the Meier case is still more open to interpretation. His 3,000 pages of notes are larded with philosophical and scientific insights seemingly beyond the knowledge of a sixth-grade dropout, yet even sympathetic investigators have been put off by the writings' sometimes petty tone. Meier's account is also partly corroborated by a number of admittedly biased witnesses, semipermanent houseguests who came to learn from Meier and remained in his home for months or years.

For instance, part of Meier's story has it that when his mind is absolutely clear, the Pleiadians can simply beam him up to their ships, much like the characters on the *Star Trek* television series. And indeed, his followers agree that Meier has vanished from his locked office in the Schmidruti farmhouse at least once. They report discovering the office locked and empty not long after Meier was seen working inside; when he reappeared hours later, three men had to break down the door to allow Meier to get back inside.

On another occasion, Meier and three helpers were up on the roof of the farm's old carriage house attaching shingles. Suddenly, Meier was nowhere to be seen. His startled companions assumed he had fallen off the roof, because there had not been time for him to climb down the single ladder unobserved. But a quick search failed to turn up his body or any sign of him. Hours later, Meier appeared, looking hale and hearty and puffing a cigarette.

Meier has also been said to enjoy demonstrating an odd power he claims the aliens gave him, gripping coins in a special way that supposedly changes them metallurgically as though they had been subjected to 1,500-degree heat. At other times—even in the middle of a discussion—he may be seized by a fit of what he has called psychic energy. His body begins to tremble, his face grows ashen, and sweat beads his brow. Without a further word he heads for his moped, apparently summoned to yet another astral encounter.

And there the evidence for and against the Meier case rests, a baffling, overwhelming body of material, none of it quite decisive. "It took us two years," UFO researcher Brit Elders told Gary Kinder, "to figure out you're never going to prove it and you're never going to disprove it. It's just there." Kinder himself came to much the same conclusion.

Whether or not Meier is a fraud, certainly other supposed UFO witnesses have been. What could motivate such a pretense? Some UFO researchers gamely suggest that contactees, perhaps including Meier, may be tempted to manufacture evidence to convince people of something that actually did happen. Frustrated by a lack of ironclad proof of their experiences, they simply create the evidence themselves. Skeptics suggest quite a number of other reasons could be adduced to explain why people would make up stories and manufacture evidence—motives ranging from sophomoric prankishness to a need for attention.

Another factor could well be the simple desire for minor celebrity or easy profit. Lecture fees, book contracts, and talk-show invitations all flow to individuals willing to discourse on their memories of alien encounters, whether sincere or faked. Even a purported UFO photograph of good quality can earn many thousands of dollars in worldwide reproduction rights from newspapers and magazines. (By way of comparison, one snapshot said to show the Loch Ness monster—a purely terrestrial entity—is estimated to have netted its owner more than $300,000.)

According to some studies, the majority of UFO pictures do not stand up to careful analysis. Ground Saucer Watch, an investigation group in Phoenix, Arizona, tests alleged UFO images with a complex process of computer enhancement. According to the organization, more than 95 percent of those it reviews are either misidentifications of natural phenomena or outright fakes. Ground Saucer Watch has rejected several Billy Meier photographs, but in the ever-contentious world of UFO research that judgment fails to settle the issue. Meier supporters criticize the Ground Saucer Watch process because it involves so much copying and alteration that they believe the photographic image is seriously distorted.

Beyond the tangled arguments over possible fraud and deception lies the problem of the honest mistake. So numerous are reports of UFOs that turn out to be weather balloons, jet aircraft, sunbeams, cloud effects, or even light reflections glancing off a windshield that UFO researchers do not even bother to tabulate them. As a child lying in a hayfield discovers castles and dragons in the clouds overhead, so the adult mind seems able to populate the heavens upon the merest optical suggestion.

In a 1988 anthology of UFO research published by the prestigious British UFO Research Association (BUFORA),

British investigator Hilary Evans offered a case in point. Sixty-five-year-old Mrs. Adams—a pseudonym—felt "called" one autumn evening in 1981 to go to her window. Outside she could see a yellow object "like two blobs of golden jelly" wobbling and pulsating in the sky; the blobs then merged, she said, and formed the shape of a cross. Greatly excited, Mrs. Adams phoned her son, who lived nearby. Young Adams saw nothing out his own window, and neither did his wife, Janette. But when Janette Adams arrived at her mother-in-law's house, the golden apparition, now constantly changing in form, appeared overhead to both of them.

According to their account, the two women watched open-mouthed as the object emitted a smoke screen to conceal itself from an aircraft that had presumably been sent to investigate. Nor was this the end of the matter. Over the next few days both women suffered severe headaches. Mrs. Adams became convinced that characters on her television screen were trying to contact her, and she lost all memory of a period of fourteen hours, seemingly experiencing the "missing time" typical of UFO abductees.

A prompt investigation by Philip Taylor and Jenny Randles of BUFORA revealed that the two women had almost certainly been haunted by nothing more than the harvest moon. The golden object, they found, had been located in the same part of the sky as the moon and had disappeared when the moon set; moreover, the very high altitude clouds noted by local weather reports would have appeared fixed in place near the moon, suggesting the supposed smoke screen. Nor could the BUFORA team find any airplane that had been sent up by the local Royal Air Force base to take a look at any odd aerial phenomena that day. As to the headaches and time loss, they could well have resulted from Mrs. Adams's highly charged emotional state.

Some suspect sightings have even occurred without a visible hint. During the 1960s, for example, German and American scientists studying the upper atmosphere launched a number of rockets designed to release luminous clouds of barium gas that were usually visible for hundreds of miles. Often, after a launch, worried citizens who saw the clouds would phone in to say they had spotted a flight of UFOs. To ward off the anticipated host of phone calls, one group of American researchers alerted the public in advance to a launch scheduled for August 16, 1966.

he tactic backfired. Intrigued by the press release, thousands of people across the American midwest stood watching the skies that night. Despite the earlier announcement, reports of UFO sightings came in from Illinois, Minnesota, Wisconsin, and other states. The only problem with the reports was that at the last minute, the barium shot had been postponed. What had everyone been watching? Nobody knows. In a final ironic twist, when the rocket finally went up a month later, not a single UFO report came in.

In a 1977 book, *Space-Time Transients and Unusual Events,* Canadian psychophysiologist Michael A. Persinger and coauthor Gyslaine Lafreniere suggested that still other mistaken UFO reports could be linked to a rare form of aerial static. Persinger's culprit is a luminous phenomenon that can be caused by physical stresses within the Earth's crust. When certain types of rock are subjected to pressure—as before an earthquake—they may give off an electrical charge, which could take the form of visible light. Shortly before an earthquake, luminous forms measuring anywhere from twenty to two hundred yards across have been seen to emanate from near the fault line. British scientist Paul Devereux posits a slightly different version of the theory, suggesting that "earthlights," as Devereux calls them, may stem from mechanical friction in the stressed rock.

Devereux has linked several outbreaks of UFO sightings to such seismically generated earthlights. The sites of UFO-like "light phenomena" reported in northwest Wales in 1904 and 1905 lay along a geologic feature known as the Mochras fault, Devereux reported, "like pearls on a thread," while faulting patterns in the Saint Brides Bay region of southwest Wales were closely associated with the locations of several sightings in 1977.

Seismic activity

UFO sightings

Concurrence of seismic activity and UFO sightings

The green areas on this map of the United States show places where, during the period of 1870 to 1960, UFO sightings occurred within a short time before an earthquake or other seismic event. Canadian psychophysiologist Michael Persinger uses such data to support his tectonic strain theory, which states that many UFO reports are actually sightings of a straightforward, if exotic, geologic phenomenon: "earthlights." Produced by physical stresses within the Earth's crust, earthlights take the form of lightning bolts, electric sparks, and fireballs, and could be mistaken for flying saucers.

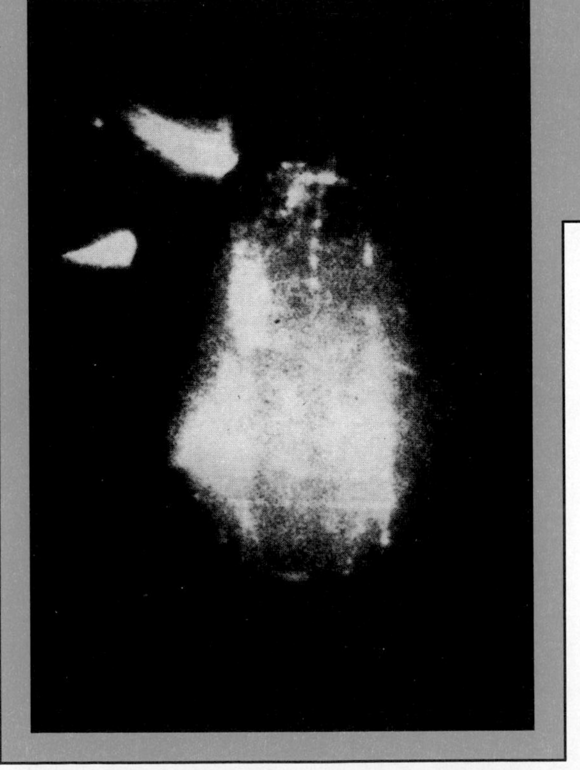

UFO-like earthlights (right) are created in a laboratory by subjecting granite to a pressure of 30,000 pounds per square inch. Electrons released by fracturing the rock make contact with air, producing luminous subatomic particles called photons.

The notion that some supposed UFOs may actually be static generated by earthquakes—a hypothesis sometimes known as the tectonic strain theory—immediately ushered in a contrapuntal chorus of affirmations and rebuttals. Perhaps no criticism was as devastating as the observation that many, and perhaps most, UFO sightings occur nowhere near major geologic faults. Persinger has attempted to answer this by suggesting that strain effects may occur more than a hundred miles from the actual quake site, while other researchers point out that tectonic strain could account for some, even if not all, UFO sightings.

Yet the tectonic strain theory remains appealing to those skeptical of UFO sightings, because it also encompasses a possible cause for mental aberrations that might

Announcing Ourselves to the Universe

"This is a present from a small, distant world, a token of our sounds, our science, our images, our music, our thoughts and our feelings." So begins former American president Jimmy Carter's address to anyone or anything in the cosmos with the wit to understand. His words join those of people from around the world on a recorded disk placed aboard two American spacecraft launched in 1977, *Voyagers 1* and *2,* in a mostly symbolic effort to send a message that other intelligent beings might someday receive and listen to.

The Voyager recording embodies a quest known as the search for extraterrestrial intelligence, or SETI. Embraced by the scientific establishment, SETI is based on the assumption that given the almost infinite number of stars in the universe, it is highly probable that life exists on other planets in other solar systems.

Five years earlier than Voyager, with the launch of the *Pioneer 10* spacecraft in 1972, experts jumped at the chance to attempt communication with alien beings: *Pioneer 10* would be the first Earth ship to enter the vastness of interstellar space.

Astronomer Carl Sagan and several of his colleagues created the message that *Pioneer 10* carried. Engraved on a six-by-nine-inch gold-plated aluminum plaque *(opposite, top),* it included diagrams of Earth's location and sketches of the spacecraft and of a human male and female. *Pioneer 11,* launched in 1973, carries an identical plaque. Each will have to travel for 30,000 years to reach the nearest star in its flight path.

In 1974, radio astronomer Frank Drake sent a message from the Arecibo, Puerto Rico, radio telescope toward a star cluster in the constellation Hercules, 25,000 light-years away. His dispatch, which intelligent beings could decode to produce an image *(right),* characterizes the chemical and genetic nature of life on Earth.

But the most informative communication from Earth thus far was that carried aboard *Voyagers 1* and *2.* It is a copper videodisk *(opposite, bottom, with cover)* that provides two hours of the sights and sounds of Earth—not only President Carter's message, but everything from whale songs and a dog's bark to greetings in fifty-five languages, samplings of the world's music, and more than one hundred photographic images. Both Voyagers left the solar system in the 1980s and began their journeys across interstellar space.

Though chances are slim that these carefully planned and heartfelt communications will reach other intelligent beings, scientists value them as gestures. Through the messages, humankind acknowledges that it might share the universe with other sapient beings.

This color bar picture represents the visual decipherment of the radio message Frank Drake beamed into space in 1974. Reading from right to left at the top of the image, which was encoded using a readily discernible mathematical system, are the numbers one through ten. Below them appear expressions for hydrogen, carbon, nitrogen, oxygen, and phosphorus, the five elements essential to life on Earth. Beneath those is a reference to the DNA molecule, the basis for life on Earth. The message's bars create pictures of DNA's double helix (blue), a human body, the Solar System (amber), and the Arecibo radio telescope (bottom), the device that sent the message.

The star-burst-like map (left) on the Pioneer 10 and 11 plaques gives Earth's location relative to pulsars, which are sources of natural radio emissions. Pulsar frequencies change over time, so the pattern shown here would help alien recipients date the message.

The Voyager disk's gold-plated cover (left) includes the Pioneer plaque's pulsar diagram explained above and drawings that tell how the disk is to be played to produce sound and pictures. The record itself is made of copper, which is resistant to corrosion in space.

make witnesses wrongly believe they had experienced an abduction. In laboratory tests, people exposed to a strong dose of electromagnetic radiation such as that attributed to tectonic strain may experience headaches, fatigue, and a state of altered consciousness. In some studies, subjects report a tingling in the spine accompanied by a sense of psychic disorder or liberating release. Some experience a temporary paralysis; others feel they part company with their bodies and travel into space. Such sensations, some theorize, could combine with sightings of earthquake-induced lights to form the basis of an abduction hallucination.

That alleged mental effect is only one of several psychological explanations for reported UFO sightings and alien abductions. Other researchers suggest, for example, that there are known medical causes for the common abductee symptom of "missing time," the lack of conscious memory of an extended period of hours or days. Multitudes of people who have never reported a UFO encounter suffer such memory lapses, which may last anywhere from hours to weeks. One cause, of course, is alcohol- or drug-induced blackout. Another is head injuries. And a third is the medical condition known as dissociative amnesia. Although more common in fiction than in fact, dissociative amnesia—a kind of complete forgetfulness—does strike otherwise healthy people subjected to unbearable distress, leaving them unaware for a time even of their own identity. Once the episode is over, a recovered victim may recall nothing of the time that has elapsed.

A number of typical abductee symptoms may also occur during epileptic seizures. Brought on by massive electric discharges of the cerebral nerve cells—a thunderstorm in the brain, as it is sometimes called—an epileptic episode sometimes begins with a feeling of profound unreality. Perceptions shift, thoughts grow confused, and—sometimes—hallucinations occur.

Strange shapes may appear to the sufferer, odd scents may seem to waft through the room, and mysterious voices may echo inside the victim's head. This effect, known to physicians as an "aura," is particularly likely when the seizure begins—as it often does—in the brain's temporal lobes, areas at the sides of the skull that are thought to control the main electric circuitry of the brain. Patients who experience an aura may be terror stricken or transported to a realm of psychic bliss. The aura may be followed by a period of automatism, in which the sufferer performs only rudimentary actions with little awareness. All of these effects have been described by purported victims of UFO abductions.

Absent considerable research funding and the full cooperation of abductees who may have moved on with their lives, ufologists find it difficult to fully refute such skeptical analyses. They can point out, however, that very few abductees are known epileptics. Indeed, most abductees appear to be perfectly healthy people with no discernible mental or neurological quirks. John Mack, a Harvard professor of psychiatry, avows that abductees with whom he has worked are "ordinary people" who relate well to others, have a strong sense of reality, and feel emotions appropriate to the experiences they describe.

The fact remains, however, that not only people with organic or psychiatric disorders hallucinate. Almost anyone can do so, given the right circumstances. In a nineteenth-century survey conducted by several early psychologists, including the well-known American philosopher William James, 3,271 people out of 27,329 respondents—nearly 12 percent—said they had experienced the sensation of being touched or spoken to by someone who was not in fact present. Such sensations are likeliest under conditions of emotional anxiety or severe bodily stress. Of course, some would contend those experiences were not hallucinatory at all, but paranormal.

Lack of sleep can also breed phantom images of almost tactile vividness. And even among the well rested, the twilight zone between wakefulness and sleep is notoriously prone to fanciful illusion. Sinking into this hypnagogic state, as psychologists call it, a sleeper first loses control of rational thought, then awareness of exact time and place. Finally, the sleeper sheds the ability to distinguish fact from fantasy. Voices may seem to call out messages or names, and faces may appear to loom out of the darkness.

A similar trancelike condition may develop with any extended repetitive act—such as driving long distances at night. Oncoming headlights dazzle, the mind grows numb, the white line spins out toward infinity, and the driver enters a state of self-induced hypnosis. The images that spring into being at such times often seem to rise up from the deepest levels of the human psyche, and their vivid, seeming reality, say those who embrace this view, could account for many of the events described by those who believe they were captured and taken to a UFO.

bductees also often report having felt fear as the aliens poked and palpitated their bodies, extracted blood and tissue, and subjected them to all the humiliations and discomforts of a searching physical examination. Some theorists, such as California UFO researcher Alvin Lawson, suggest that the alleged abductees are reliving the moment of their own births. Or maybe they are reexperiencing a more recent trauma—a stay in a hospital, perhaps, with all its attendant anxieties. But every human being was born, and many have been treated in hospitals. Assuming that alleged alien abductions are simply fantasies based on those commonplace experiences, skeptics must explain what specific mental state would cause a few people to develop such beliefs while most do not.

One force that may be at work was suggested by Carl Jung in his groundbreaking 1958 work, *Flying Saucers: A Modern Myth of Things Seen in the Sky.* Central to Jung's psychological theories was the notion of cultural archetypes, primal images that Jung believed underlie all human awareness and that serve as the foundation stones of cultural and religious belief. The images emerge, Jung thought, from what he called the collective unconscious, a kind of stored reservoir of human beliefs. While the images' external trappings may differ from age to age, at the deepest level they are fundamentally the same.

According to Jung, the most basic archetype is the circle—complete, perfect, without beginning or end. In the form of the Hindu mandala, a circular emblem often containing pictures of various deities, it represents the totality of the universe. Pictured in the ancient symbol of the *ouroborus,* a snake biting its tail, it symbolizes eternity. "There is an old saying," Jung wrote, "that God is a circle whose center is everywhere and the circumference nowhere." In many instances, he pointed out, the divine circle is envisioned as radiating fire and light. Most flying saucers, true to their name, are said to be circular in shape—and often glowing.

This is not mere coincidence, Jung suggested. He published his *Saucer* essay at the height of the Cold War, in the decade of nuclear bomb shelters and one year after the Soviet Union launched Sputnik. "In the threatening situation of the world today," he wrote, "the projection-creating fantasy soars beyond the realm of earthly organizations and powers into the heavens, into interstellar space, where the rulers of human fate, the gods, once had their abode in the planets." Alien spacecraft could be the modern equivalent of Hindu mandalas, according to Jung, dusted off and rolled to center stage to assuage our terror of nuclear war.

Since then, Jungian enthusiasts have expanded his proposition from the mere sighting of heavenly UFOs to reports of human abductions by their crews. Hilary Evans, for instance, has suggested that abduction stories are a form of "hallucinated psychodrama," in which the supposed victim attempts to exorcise some inner personal crisis. Whereas in past ages the psychodrama would involve witches, angels, and demons, Evans says, it may now be expressed "in terms of the prevailing myth of our time, the myth of extraterrestrial visitation."

Although cloaked in different terms in different ages, stories about celestial visitors are certainly nothing new. Many human cultures offer tales of beings from the heavens who descend to Earth for the betterment or discomfiture of humankind. Sometimes the visitors come enveloped in clouds of glory, but a few are said to have arrived in what could be construed as actual spacecraft. The skyborne vision of the Old Testament prophet Ezekiel, which included spinning wheels and flashing fires, is often cited by ufologists as a picture of a prototype UFO, as are the fiery chariot in which Elijah rode to heaven and the radiant cloud in which he is said to have reappeared, centuries later, to speak with Christ on a high mountain.

For generations during the Middle Ages, devout Christians spotted flaming crosses in the sky and flying archangels with their shields all aglow. Even secular visitors would sometimes arrive from the air, according to popular belief. As long ago as AD 840, the archbishop of Lyons drafted a stern letter condemning what he considered the folly and stupidity of his parishioners, who claimed that "cloud ships" were swooping down from a sky kingdom named Magonia and making off with unsuspecting locals.

Students of folklore can point to any number of similar reports. One example cited by French ufologist Jacques Vallee is an Algonquin Indian story of a hunter who spies twelve beautiful maidens wafting down from heaven in a large basket. He catches one and makes her his wife. They have a son. But the woman grows homesick for her celestial abode. She weaves herself another basket, climbs in with her child, and takes off. Later, she returns to Earth to fetch her Algonquin husband. On the couple's return to the sky, the heavenly elders lay on a big feast that includes samples of the earthly game that the husband, by special request, has brought with him.

According to Vallee, the Algonquin legend contains all the elements of a top-flight UFO abduction account. Besides the disklike basket from a distant star, there is the element of sexual union between human and alien. Even the star elders' desire to savor terrestrial wildlife mirrors modern reports of animal mutilation and other alleged specimen taking by starship visitors.

Vallee has pointed to a number of other folklore elements equally reminiscent of abduction reports. Medieval witches, for example, were said to return from their riotous sabbaths with telltale bodily scars as evidence of intimate relations with the devil, scars reminiscent of those that ab-

Abduction as Psychodrama

Many people who believe they have been victims of alien abduction suffer lasting psychological distress as a result of their terrifying experiences. Sights and sounds of the purported encounter haunt them in grotesque flights of memory, and an inner voice bombards them with answerless questions: Did the abduction really occur? Why did they take *me?* Did I imagine it all? Am I crazy?

"1000 Airplanes on the Roof," a contem[illegible]y American drama accompanied by music, explor[illegible]s torment. Fascinated by the psychological dimension of alien encounters, the work's creators, playwright David Henry Hwang, composer Philip Glass, and scenic designer Jerome Sirlin, produced a disturbing and highly accurate representation of postabduction anguish.

The single onstage character in the drama is called M. Before his abduction he was a successful lawyer who lived in a converted farmhouse with his wife and family. Now, living alone in Manhattan and holding a menial job, M. *(right)* tells of his abduction: One night he was suddenly taken captive and transported to a spaceship, where alien beings performed medical examinations upon him. Later they released him, warning him to forget the entire occurrence.

But he cannot forget. A sound like that of 1,000 airplanes pounds in his head—thus the drama's title. Fragments of the abduction scene filter in and out of his consciousness, and he thinks he is going insa[illegible]e wants to tell someone, perhaps the woman he is out with on a first date, his nightmarish secret, but he fears being laughed at or taken for a lunatic. The audience shares in M.'s horror as tension-inspiring music plays and holographic projections of dreadful alien faces appear onstage behind him.

Reviewers of the 1988 production, who generally praised it, saw it as a metaphor for loneliness and loss of control over one's destiny. For abduction victims, the drama would play more like a documentary.

ductees trace to their own unwelcome shipboard examinations. Incidents of missing time are also found in the world's mythologies; perhaps the most familiar American example is that of Rip Van Winkle, in Washington Irving's folklore-inspired story by that name.

Even minor details of the abduction experience have parallels in folklore. Vallee notes that a number of alleged abductees recall seeing a luminous crystal during the time they were held captive. The stone may have been incorporated into the aliens' uniforms, installed within their spaceship, or located on the rarely glimpsed home planet. Similar shining stones, he says, were standard equipment for certain kinds of French pixies.

To call something a myth, however, does not prove that it has no basis in experience. Vallee himself believes there must be some basis for the recurrent tales of UFOs—as did Jung. In his 1958 treatise, Jung acknowledged that some UFOs could well be real physical entities—particularly when they showed up on Air Force radar screens, as in a well-known 1952 sighting near Washington, D.C. In that case, one of many similar incidents around the world that year, radar operators noticed seven slow-moving objects on their screens, while controllers at a nearby airport reported seeing mysterious airborne lights. "Either psychic projections throw back a radar echo," Jung commented wryly about such events, "or else the appearance of real objects affords an opportunity for mythological projections." He seemed to favor the second option.

And that is the classic explanation of the UFO puzzle. Ever since the advent of modern rocketry during the Second World War, some observers have suggested that UFOs are spacecraft piloted by astronauts from beyond the planet Earth. If so, the ships need all the speed and agility they are said to exhibit in Earth's skies. Superrapid transit would be essential for any extraterrestrial astronaut who wanted to visit Earth, simply because of the distances involved. Some 93 million miles of vacant space separate the Earth from the Sun; another 3.57 billion miles, on average, stretch between Earth's orbit and that of Pluto, a tiny frozen rock frosted with methane ice that usually lies at the outer limit of the known solar system. (In 1979, Pluto's eccentric orbit brought it inside that of Neptune, so that for two decades Neptune became the outermost known planet.)

In all this expanse, only the planet Earth provides a physical environment that could spawn and support intelligent life as we can imagine it. Frigid Mars has little atmosphere and no liquid water to speak of, and NASA's Viking landers found no microorganisms at all in its soil—a finding that essentially eliminates the possibility of life there. Venus, with a surface temperature of nearly 900 degrees Fahrenheit and permanent sulfuric acid cloud cover, is still less hospitable to living organisms, and sunbaked Mercury, the innermost planet of the solar system, resembles nothing so much as Earth's own lifeless, cratered moon.

With the exception of rocky Pluto, the outer planets are gigantic globes of swirling gases seemingly inimical to life, with highly pressured oceans tens of thousands of miles deep inside. Some are circled by large moons with wispy atmospheres, but these satellites seem too cold for life. Although at least one eager theorist has proposed still another, as-yet-undiscovered planet *(pages 89-97)*, most astronomers believe that a world beyond Pluto would be colder yet, and even less likely to have developed inhabitants. Logically, it seems, if any alien vehicles visit the Earth, they almost certainly come from somewhere in the vast reaches beyond our solar system.

Most scientists agree that intelligent life might indeed exist elsewhere. Earth's sun is part of a conglomeration of about 100 billion stars called the Milky Way galaxy, a vast spiral disk some 100,000 light-years across. (A light-year, about 6 trillion miles, is defined as the distance that light can travel in a year). Current theory suggests that some of the galaxy's many stars are orbited by planets, and that some fraction of those planets are, like Earth, capable of supporting life and even intelligence.

To be sure, no one has yet seen an actual planet beyond our own system, nor is a sighting likely in the near future, because such planets must be so distant they would appear infinitesimally small. For some decades, however, eager astronomers have detected indirect evidence of planets or planet formation around a number of stars.

In 1991, for example, astronomers at the Arecibo radio telescope in Puerto Rico picked up signals that seemed to suggest a planetary complex. The pulses emanate from a tiny neutron star known as PSR 1257+12, and they indicate that two, and possibly three, small objects revolve around the massive, collapsed star at the center of the system. Given their proximity to a tremendously condensed star thought to have formed from a kind of stellar explosion millions of years ago, however, these miniature planets are unlikely settings for alien life. Furthermore, they swing through space some 7,800 trillion miles from Earth, a distance equivalent to 1,300 light-years. For scientifically minded UFO theorists, such distances pose a problem.

According to Albert Einstein's theory of relativity, any object increases in mass when it goes faster. Since this strange phenomenon is unnoticeable at ordinary speeds, the notion goes against common sense. Yet when particles are deliberately accelerated to very high speeds in a laboratory setting, they do indeed increase in mass, just as the theory predicts. Einstein's equations also show that when a spaceship or any other object approaches the speed of light, approximately 186,000 miles per second, its mass becomes almost infinite. Mathematically speaking, a truly infinite mass would be impossible to accelerate, so no physical object can quite reach the speed of light. That cosmic speed limit is a considerable obstacle for space travelers. Moving just below the speed of light, a starship would take more than 1,300 years to reach Earth from the PSR 1257+12 neutron star system.

Certainly, there are other, perhaps somewhat closer, celestial orbs that could serve as launching pads for spacefaring species. Sebastian von Hoerner of New Mexico's National Radio Astronomy Observatory suggests that perhaps 2 percent of all stars are orbited by "a planet fulfilling all known conditions needed to develop life similar to ours." Similarly, in his 1979 book *Extraterrestrial Civilizations,* the late author and biochemist Isaac Asimov speculated that as many as 390 million of the galaxy's stars are circled by planets now populated or formerly populated by intelligent species.

Even so, the distances between inhabited planets are vast, which is one reason that Asimov and other skeptics have long dismissed all UFO reports out of hand. Asimov's most optimistic estimates were based on the assumption that civilizations endure for billions of years rather than going up in the catastrophic smoke of runaway pollution, rampant population growth, or nuclear holocaust. Given such long-lived cultures, his computations suggested that on the average, most galactic technological civilizations would be forty light-years apart. If technical cultures endure only a few thousand years, or even a few million, the distances between civilizations active at the same period in galactic history would be correspondingly greater.

Asimov's numbers correspond with figures cited by the UFO community. BUFORA researcher John Prytz, for example, suggests that the nearest technological civilization to our own is likely to be more than 1,000 light-years away, requiring a minimum of more than 1,000 years to reach through conventional space. Even that scenario supposes travel speeds very close to the speed of light, in itself a Herculean feat.

On the other hand, say some ufologists, perhaps the aliens have somehow managed to sidestep the mathematical limits of Einsteinian relativity. Perhaps their pilots have found alternative pathways in a theoretical hyperspace, moving through time warps, or dodging in and out of celestial black holes—both strategies which have been suggested as options. Others have optimistically, if vaguely, proposed that a broader and more permissive theory of physics could be formulated that encompasses the laws propounded by Albert Einstein just as Einstein's work included—but went far beyond—the postulates of Isaac Newton.

John Shepherd stands amid the welter of electronic gear of his homemade UFO communications and detection center in his house in Michigan. For several hours each morning and evening he tries to "appeal directly to the heart, mind, and soul" of extraterrestrials by sending out music and repeating sequences of musical tones. His instruments also monitor the Earth's magnetic field for changes that might be attributable to UFOs. If such a fluctuation occurred, Shepherd says that he would send a message of greeting and peace to the visitors.

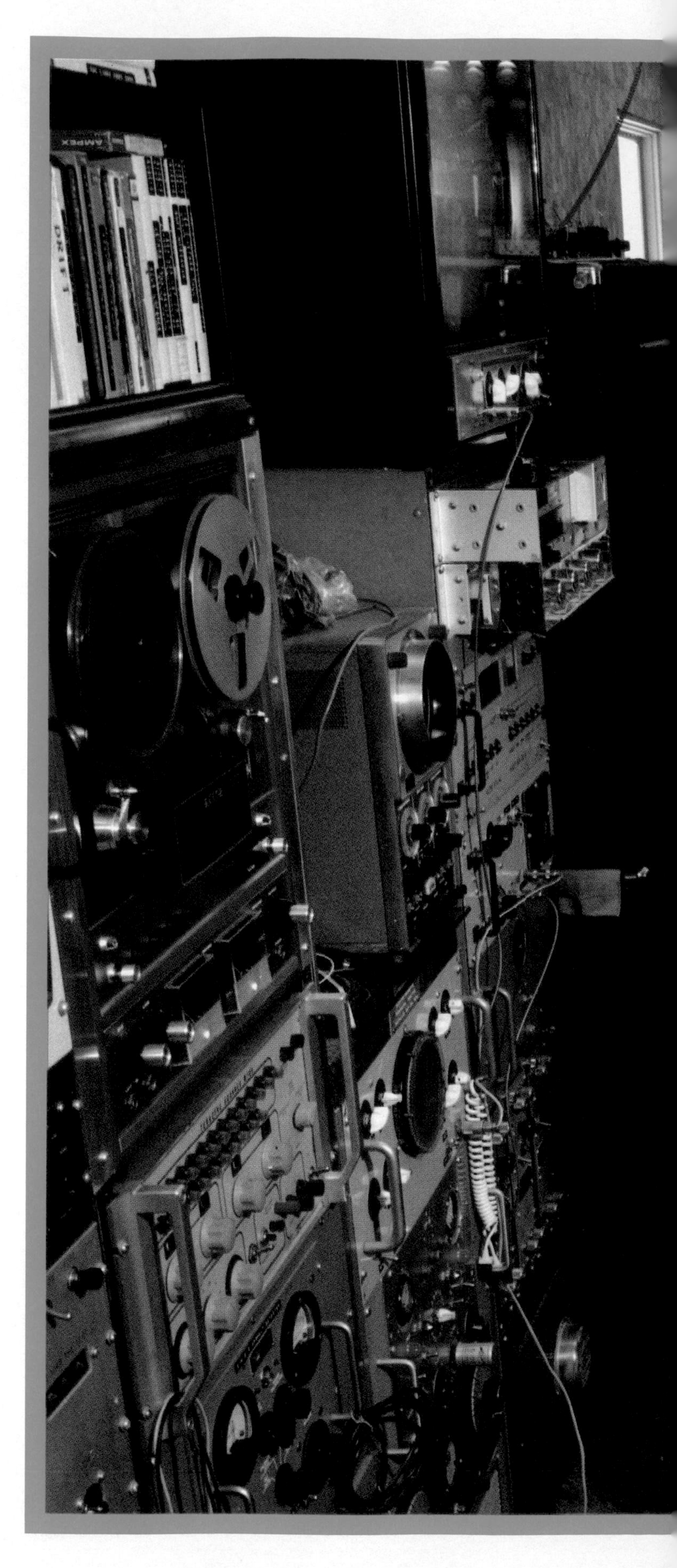

Nor is such speculation limited to UFO enthusiasts. In May 1988, top-ranking cosmologists Kip Thorne and Michael Morris amused themselves by speculating in print just how black holes could conceivably transport travelers through space or time. Their proposed system involved two black holes at great distances. Black holes, themselves hypothetical objects that have yet to be definitely proven to exist, are immensely dense bodies whose gravity distorts the ordinary properties of time and space. Linked together in another dimension, two widely separated black holes could theoretically form a tube or tunnel known as a wormhole, a mathematical construct first proposed in 1916.

Traditionally, wormholes have been seen as fatal to any traveler, since they would close upon and crush anything that entered their dangerous domain. But in their article, Thorne and Morris suggested that a bizarre type of material known as "exotic matter" could be made to hold a wormhole open long enough to grant a space voyager safe passage. Playfully, they envisioned such a traveler journeying through time or space in the elapsed time it might take an Earthling to start across a road.

Assuming that alien astronauts have used that strategy—or some other technology as yet unimagined by humankind—in order to visit the Earth, one gigantic question remains: Why do they bother? All kinds of reasons have been put forward, although guesses about alien motives may reveal more about human thinking than about alien psychology. UFO crews have been said to be motivated by everything from scientific curiosity to a predatory urge to acquire distant space colonies. One suggestion paints a scenario in which the aliens have exhausted the physical resources of their native planet and are now prospecting other worlds for vital minerals, energy, and food supplies. In 1988 ufologist Richard Hall put forward the idea, in all seriousness, that the aliens are after Persian Gulf oil.

John Prytz, Michael Swords, and others have suggested that the aliens are migrating into space because of population pressures at home—the Westward Ho approach to space exploration. Survival is a basic urge of all life, after

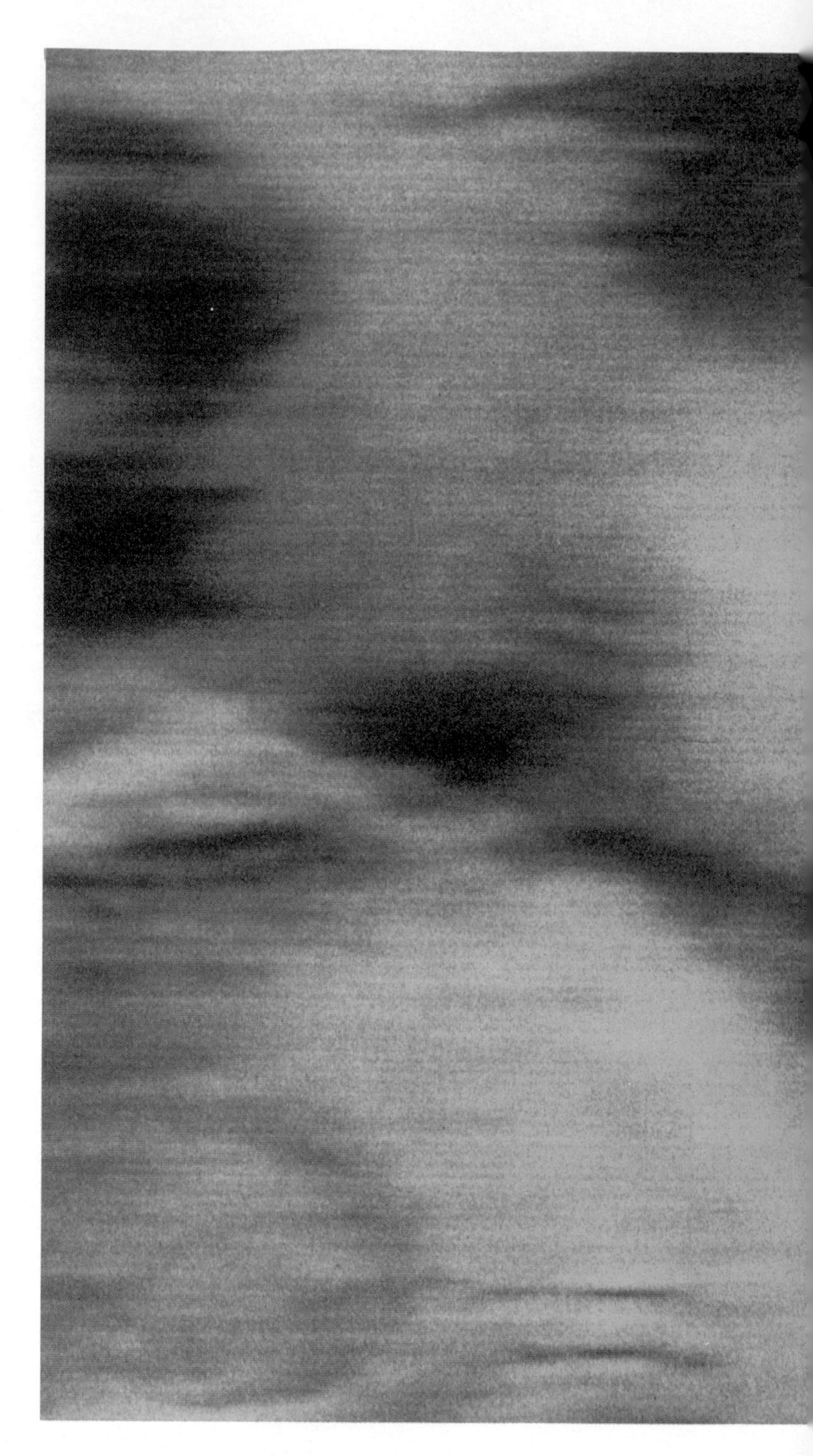

all, yet every planet is faced with ultimate destruction when its star ages and dies. Any civilization that manages to establish colonies on other worlds increases its chances of indefinite survival.

A more sinister possibility has been raised by Budd Hopkins, who has done more to call public attention to the dark side of alien intrusion than any other ufologist. Although he is best known for his theory that aliens are experimenting on human subjects in order to create a new, hybrid interstellar species, Hopkins also believes that the aliens have psychological needs. The aliens, Hopkins surmises, are somehow deficient in basic vitality. The human psyche, in all its emotional strength and complexity, fascinates them. "Abductee accounts suggest that the aliens are interested in acquiring, or at least understanding, the basic human emotional spectrum," Hopkins asserts.

ertainly, Hopkins's picture of lonely star travelers irresistibly drawn to manipulate human destiny remains one of the few theories built upon the details of alien abduction reports, rather than simple guesswork. Yet the puzzle of how aliens could come such a long way, where they stay between visitations, and what could make them want to travel so very far makes any version of the extraterrestrial visitation theory difficult for some ufologists to accept.

Frustrated by the seemingly insurmountable barrier presented by the speed of light, scientists interested in the UFO phenomenon have come up with what they see as alternative origins for the alien visitors. According to Einstein's theories, for example, any object that accelerates toward the speed of light undergoes a bizarre transformation. Not only does its mass increase, but its volume shrinks and its own rate of time slows toward zero. Hypothetically, if an object could reach the light-speed barrier, time would have stopped altogether, and the object—now infinitely small, and infinitely heavy—could be said to vanish. In theory, at least, it drops out of our familar space-time universe. And where does it go? The answer, according to ufologist Michael Swords, may be that it pops up in another universe, composed of other dimensions.

For those who believe UFOs are indeed crewed by interstellar travelers, this might provide a cosmic shortcut for alien space pilots, who could speed through the barrier of the present universe, then emerge an instant later in the vicinity of Earth. But Swords speculates there are other possibilities as well. If time dwindles to zero at the critical moment, it could show up on the other side as a mathematical negative. In effect, it would run backward. This suggests to

A Martian Mystery

While its *Phobos 2* spacecraft was orbiting Mars on March 27, 1989, Soviet mission control lost touch with the satellite under unexplained circumstances. Official comment on the incident was minimal, but it raised the antennae of UFO enthusiasts around the world. There appeared to be enough information to suspect that world leaders were hiding a disturbing possibility: The Soviet probe may have been attacked by an alien craft.

Phobos 2 lifted off in mid-1988 to study Mars and its moon Phobos, namesake of the craft. By the next March, it had sent many photos of Mars and the lumpy, elliptical Phobos and was scheduled to send more.

At 3:59 p.m. Moscow time, the craft's signal abruptly failed. Nothing more was heard from *Phobos 2* until 8:50 p.m., when weak contact was reestablished for thirteen minutes, after which the signal ceased permanently. Engineers later reported that *Phobos 2* had gone spinning out of control and that the probable culprit was either a computer malfunction or, more mysteriously, a collision with an unidentified object.

Scientists pored over the last bits of film the probe sent back to Earth, but these produced more questions than answers. According to Soviet television news, the photographs recorded by *Phobos 2*'s cameras, which were aimed at Mars at the time, revealed a dark elliptical shape between the satellite and the planet's surface *(left)*. The final photographic image, which has yet to be made public, allegedly is stranger still. As a British scientist said, the Soviets apparently saw something in it "which should not be there."

Since no other official information has been released, the cause of *Phobos 2*'s demise is open to speculation. American author Zecharia Sitchin, whose theory of the alien heritage of humankind appears on pages 89-97, is joined by other members of the UFO community in the opinion that the dark object in the photos is an alien spacecraft. He thinks the secret final photograph shows the first instance on record of real star wars: an alien craft attacking and knocking out the interloper *Phobos 2*.

Swords that alien astronauts may be travelers from the future, who have come back to present-day Earth in order to recapture their own past, and perhaps to change the course of history for the better.

Time travel, whether achieved through that mechanism or another yet to be imagined, is generally thought to be an impossibility, in part because it creates any number of logical paradoxes. The most frequently cited is the case of a futuristic lunatic who comes back and murders his own grandfather. Would this rash act be a kind of patriarchal suicide, which would prevent the lunatic's own birth? If so, how could he come back in the first place? The answer, according to Swords, could lie in the concept that any number of universes exist simultaneously, generally invisible one to the other, but occasionally interpenetrating. Thus, the lunatic could exist in one universe but not in the others.

Still another explanation of the UFO phenomenon—half-scientific, half-metaphysical—is that of Jacques Vallee, the same writer who popularized Jung's folklore hypothesis in the 1960s. In his writings, Vallee has suggested that aliens are not from other worlds in our universe but from an alternate reality that exists parallel to our own. Vallee, who began his professional career as an astrophysicist at the Paris Observatory, has published a number of books on UFOs, a topic which has come to be his life's work. Well known in the field and relatively publicity shy, Vallee stands apart from other UFO researchers. His reputation is such that Vallee is said to have served as the basis for the cosmopolitan French ufologist in the 1977 film *Close Encounters of the Third Kind.* Played by French director François Truffaut, the character was notable for his open mind and all-encompassing theories on alien visitation.

In his first books, published in the early 1960s, Vallee made the scientific case for extraterrestrial visitation. By 1969, when he published *Passport to Magonia,* Vallee was more interested in the folklore hypothesis. (The title is a reference to the Magonian cloud ships that so troubled the archbishop of Lyons a millennium ago.) Ten years later came *Messengers of Deception,* a study of West Coast UFO cults in the United States. Then, in 1988, Vallee published *Dimensions,* a further development in his views which attributed the UFO phenomenon to attempts to modify the collective unconscious of human beliefs first posited by Carl Jung. UFOs are shown to human beings in a deliberately tantalizing fashion, he asserted, and they are drawn from a realm beyond space and time. Although "UFOs are real physical objects," he wrote, "they are not necessarily extraterrestrial spacecraft. To put it bluntly the extraterrestrial theory is not strange enough to explain the facts." Instead,

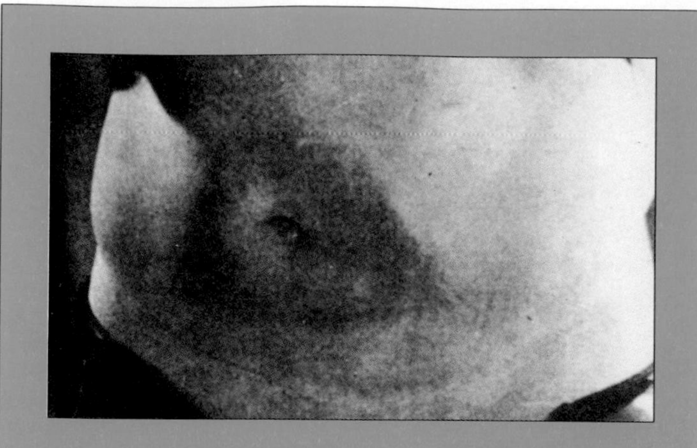

An Icon Shared by Aliens and Earthlings

For more than a dozen years before his alleged abduction by aliens in 1985, author Whitley Strieber had been fascinated by the mystical idea of the triad. Through study of metaphysical writings he had learned to view three-part entities as symbols of perfect earthly wholeness, the penultimate station on the road to transcendence and enlightenment. Because of this belief, he felt an eerie sense of connectedness with his alien captors when he recognized that to them, too, the triad was a potent symbol.

In his 1987 bestseller, *Communion,* Strieber wrote that the aliens revealed the importance they gave to the triad in many ways, one of which was by marking the bodies of abductees. Strieber himself, for instance, returned from his abduction with a triangular mark etched upon his arm. Similarly, a French physician and his son developed persistent triangular rashes around their navels after an alien encounter. The doctor's strange affliction is pictured above.

Strieber also saw triune symbolism in the reports of multiple-abductee Betty Andreasson *(page 30).* She described having been led through an alien realm, where she encountered a large eaglelike bird that Strieber associates with another tripartite symbol, the Sphinx. In Greek mythology the Sphinx asks Oedipus a three-part riddle: What has the strength of a bull, the courage of a lion, and the intelligence of a man? When Oedipus correctly replies that it is the Sphinx itself in its triple nature, the Sphinx transmutates into an eagle.

Strieber writes of other purported alien connections to the triad. The beings often appear in threes, they shine three-pointed lights, and they wear triangular emblems. All these circumstances have convinced Strieber that the extraterrestrials are seeking to convey the meaning of the triad to Earthlings. Perhaps, he speculates, the aliens are even inviting the inhabitants of Earth to couple with them to form a third entity, more perfect than either of its antecedents.

he wrote, "I believe that the UFO phenomenon represents evidence for dimensions beyond spacetime; the UFOs may not come from ordinary space, but from a multiverse which is all around us."

Modern science, Vallee argues, is increasingly open to the possibility of alternate universes. To nuclear physicists toiling in the subatomic microrealm of quarks, muons, and neutrinos, and to astrophysicists contemplating such vast imponderables as the birth and death of the cosmos itself, the solid, three-dimensional world of daily life is seen as a tiny fraction of overall reality. At the beginning of this century, Albert Einstein added time as a fourth dimension. Since then, theoretical scientists attempting to describe the structure and behavior of our universe have devised mathematical formulas that call for even more dimensions, some of which are "folded inside" others.

Although he cloaks the multiverse in the language of modern physics, Vallee does not see it as a strictly mundane physical phenomenon. Instead, he speculates that the multiverse might represent a separate "plane" similar to the one he believes is inhabited by the spirits of the dead. UFOs, like ghosts or other spirits, would thus be resistant to skeptical observation or precise measurement—yet, he says, they would still exist. UFOs "are constructed both as physical craft (a fact which has long appeared to me undeniable)," he wrote, "and as psychic devices whose exact properties remain to be defined."

The reason for such puzzling visitations from the multiverse is still unclear to Vallee, but he speculates that it might have something to do with a purposeful extraterrestrial control system that guides the course of human evolution, either physically or spiritually. "When I speak of a spiritual control system," he explained in *Dimensions,* "I do not mean that some higher supercivilization has locked us inside the constraints of a space-bound jail, closely monitored by entities we might call angels or demons." Instead, he argued, the control system is using human curiosity about the UFO phenomenon to forward its goal of creating long-term changes in Earth's culture.

By the very nature of this scenario, human beings could understand the goals of this gigantic experiment no better than a rat comprehends the purpose of a psychologist's maze. And there is no reason to think the system's ultimate purpose is benign, Vallee cautioned. "There is a strange urge in my mind: I would like to stop behaving as if I am a rat pressing levers—even if I have to give up the cheese and go hungry," he wrote, with the cheese representing further investigation into UFO anomalies. "I would like to step outside the conditioning maze and see what makes it tick," he continued. "I wonder what I would find. Perhaps a terrible superhuman monstrosity the very contemplation of which would make a person insane? Perhaps a solemn gathering of sages? Or the maddening simplicity of unattended clockwork?"

In his next book, *Confrontations,* Vallee suggested a possible answer to that question. By now, he had concluded that the experimenters were far from a "solemn gathering of sages"; instead, he saw them as potentially malevolent intruders to be feared and, if possible, resisted.

allee's complex mixture of exotic physics, behavioral psychology, and psychic theory has been influential enough that other UFO researchers are reinterpreting the matter as a truly psychic phenomenon best suited for parapsychological study. Veteran ufologist Ann Druffel took just that approach when she and colleague Scott Rogo examined the so-called Tujunga Canyon incidents, a series of reported abductions linked to the very early Sara Shaw case in southern California *(pages 22-23).* In the years that followed the Tujunga Canyon incidents, Sara Shaw's companion Jan Whitley and three other women, all of whom were interested in studies of metaphysics and psychic phenomena, were purportedly contacted again and again—perhaps, says Druffel, because of their interest in the paranormal.

Three years after she and Sara Shaw (the names are pseudonyms chosen by Druffel and Rogo) were first abducted, according to Jan's hypnotically released memories, Jan

In this German print, a man breaks through the cosmos and peeks into the endless abode of God. There, gleaming, UFO-like disks and a wheel inside a wheel—a reference to the otherworldly vision of the biblical Ezekiel—hang among pristine clouds. The picture could be considered symbolic of the parallel universe theory, which states that a different reality exists alongside the known realm. Some people believe that aliens come from this unfathomable other dimension.

and a friend named Emily had another alien encounter at a rest stop along a California highway. The two women were tired from driving and had pulled off to get some rest. Instead, their attention was caught by a strange white light that floated down from the trees. Both women then made mental contact with humanoids inside it.

For years after this curious but seemingly innocuous incident, both Jan and Emily say they were visited at night by unseen presences, perhaps related to the light's humanoid crew, who called out pleadingly and attempted to force them away to an alien world. Jan's experiences of this type were particularly unsettling. As she lay in bed in a state of paralysis, she drew on all her reserves of psychic will to prevent being kidnapped. Finally, by intense concentration, she would move a finger or a toe, causing the spell to break.

In about 1970, the aliens reportedly shifted their attention to another friend of Jan's, Lori Briggs. The first contact, according to Lori's recollections, came when a humanoid figure with intensely glowing eyes invaded Lori's bedroom, held her paralyzed, and implanted some sort of receptor device in her spine. Then, in 1975, Lori and a friend named Jo were pulled out of their house by an irresistible force and drawn into a spaceship.

In later hypnotherapy sessions, Lori recalled coming to rest on a pink, stonelike table inside a room paneled with mirrors that resembled lizard scales. Reflected in these multifaceted surfaces she saw her blood vessels pumping, their contents glowing like iridescent rivers. But what followed was not the customary semiscientific procedure. Instead, the aliens told Lori that they abducted her and other special individuals like her as a means of giving their own luminous but insubstantial bodies the power and permanence of solid flesh. In essence, the supposed aliens were disembodied spirits, eager to enter the material world.

To Druffel, who conducted the hypnotic regressions on Lori, Sara, and Emily, such stories bear all the earmarks of a classic psychic experience. Citing Vallee's work and his arguments that UFOs are too "strange" to be explained by ordinary space travel, she argued instead that the aliens were strictly psychic entities akin to demons or angels. "Human interaction with UFO entities and close encounters with their craft," she wrote, "takes place in an altered state of reality. In other words, UFO abductions do not seem to be physical reality in the normal sense of the word, possibly because the UFOs themselves are craft from other space-time continuums."

Brad Steiger, an author who has spent more than thirty years investigating hauntings, psychic powers, and other paranormal phenomena, also classes UFOs with most other supernatural phenomena. "In my opinion," he wrote in 1981, "the sightings of UFOs and the appearance of elves, angels, and other archetypal images throughout the world signify that we humans are part of a larger community of intelligences, a far more complex hierarchy of powers and principalities, a potentially richer kingdom of interrelated species—both physical and nonphysical—than we may have dreamed of."

If the California women were understandably reluctant subjects of such supernatural alien contacts, large numbers of people with an interest in paranormal phenomena eagerly embrace the experience. Ever since the early 1950s, certain individuals have claimed to act as conduits between ordinary Earthlings and denizens of the alien realm. Like spiritualistic mediums, most of these channelers believe they are communicating the thoughts of priestlike mentors. The difference is that the channelers' contacts are said to be physical beings voyaging on actual UFOs at the time of contact.

Like Billy Meier, with his thousands of pages of transcripts of Pleiadian thought, UFO channelers have given rise to a considerable body of literature. The Portals of Light organization, for example, which is based in the New Age center of Sedona, Arizona, has published a number of anthologies of messages channeled by aliens. In the 1989 book *Conclave: Meeting of the Ones,* a channeler known as Tuieta offers a rich volume of the information "received" in late 1988. The material began to arrive, Tuieta reported, shortly after 11:15 a.m. on December 16, when a friend pointed out a cloud in the sky. "When I saw 'the cloud,' I felt such an intensity of energy that I could not describe it," Tuieta wrote. "There were many small 'clouds' of the same shape on either side of the big 'cloud.' All together the combination seemed to fill half the sky." Somehow, "I *knew* this was the command ship that was used by Cuptan Fetogia," the account continued, a reference to one of the aliens

Claude Vorilhon, shown here before delivering a lecture in California, claims he learned the secrets of the universe from aliens who visited him in 1973 near his home in the south of France. The extraterrestrials renamed him Raël and revealed to him the unlikely symbol that he now wears—a swastika inside a Star of David, which means, he reports, "as above, so below." Raël now travels the globe preaching the aliens' message: world government, pacifism, nuclear disarmament, and economic reform. He has thousands of followers worldwide.

Tuieta sometimes channeled. "I began to cry with a sense of joy and sadness all at the same time. Truly, something of great significance caused the Elder Statesman to come into the ethers of Earth!"

As revealed in the rest of the book, which is a literal transcription of alien messages that Tuieta acknowledges "does not meet the literary standards of man," Fetogia had apparently come to alert enlightened Earth dwellers of a sweeping historical change. Warning of a "great inpouring of cosmic and angelic energies," Fetogia urged Tuieta to tell humanity to "expect the unexpected" and to "accept the acceleration of your vibrational pattern." He urged human beings not to fight this change, for resistance "could be quite detrimental to your health mode."

According to Tuieta, the same change became the subject of a later conference held on board Fetogia's blue-carpeted flagship. At the meeting, which Tuieta is said to have attended in spirit, Fetogia and other alien "commanders" discussed among other topics the matter of manipulating Earth's crustal plates. To avoid too much surface disruption, they were said to be working mainly with changes on the seafloor, in cooperation with dolphins, the "brothers and sisters of the waters."

Publishing is not the only way in which channelers and other contactees communicate. Many prefer to gather in groups to share the teachings of their "aliens." Such quasi-religious cults are dedicated to receiving and disseminating the flow of astral wisdom, which often revolves around a coming "great change" like the one Tuieta foretells. Although the tenor of their message may closely resemble that of Tuieta, however, the names and details—and thus the explanation for the UFO phenomenon—vary from group to group.

A number of these UFO organizations are decades old and well established. The Urantia Foundation of Chicago, for instance, first incorporated in 1950 for the purpose of promulgating the revelations of *The Urantia Book,* which had supposedly been received from alien intelligences. More than four decades later, the foundation was still thriving. By the early 1990s, according to an account in the *Wall Street Journal,* the United States had some twenty organizations that were claiming to channel astral messages. Most made their headquarters on the West Coast, but one in Florida claimed 1,500 members.

Some cult members and others also believe they share a special distinction that few ordinary UFO researchers can claim. Rather than channeling the thoughts of aliens—or simply meeting aliens during an abduction experience—some people assert that they are aliens themselves, or descendants of alien ancestors, who have now been given license to come forward.

The world is apparently full of such star people, a term used by Brad Steiger, who believes himself and his wife to be among the alien-descended elite. According to Steiger, star people are often unaware of their special identity until late in life, but they share certain identifiable characteristics. Most are heavy lidded, with "bedroom eyes," charming and intelligent, and good confidantes. They tend to have unusual blood types, extra or transitional vertebrae, low body temperatures and blood pressures, and—in some cases—misplaced ribs.

Whether or not they realize that they are of alien blood, most star people have, like Billy Meier, experienced close encounters with "what they believe to be angels, elves, fairies, masters, teachers, or openly declared UFO intelligences" throughout their lives, Steiger has written. Steiger's own first alien meeting, he says, occurred when he was just five years old. During these sessions, which like the channelers' are signs of a great change in store, the alien contacts tend to deliver stern warnings on the depth of human folly—recent topics include such serious matters as nuclear proliferation and environmental waste—and generally prepare their earthly pupils for the "approaching time of transition, the great cleansing, which the entire species must endure in order to attain a higher state of consciousness," as Steiger puts it.

Such themes of alien guidance and immense future change pervade psychic literature on UFOs. But for all the mystical claims of UFO channelers, UFO wisdom societies, and quondam star people, most conventional UFO researchers discount their elaborate claims, which tend to lack any form of physical evidence.

Yet no ufologist doubts that the research into every aspect of this complex phenomenon should continue, for the UFO puzzle shows no signs of imminent solution. The key for those who would unravel the mystery, wrote *Journal of UFO Studies* editor Michael Swords in 1990, is to approach each investigation in a dispassionate, scientific manner, unclouded by prejudice or emotional bias. "Not only is it worth keeping an open mind about these things. It is, in this age of newfound scientific humility," Swords wrote, "imperative to do so."

When Art and UFOs Meet

Soon after he reportedly spotted a disk-shaped UFO in the skies above Cape Cod in 1964, New York artist Budd Hopkins found that the experience was having an effect on his life and art. "A memory trace of the UFO's appearance," he wrote, "began to infiltrate my paintings in the form of a dark, centralizing circle." Hopkins set out to inquire whether other artists' works had been influenced—directly or indirectly—by UFOs. He concluded that by the mid-1950s, media coverage of "flying saucers" had exposed the world to popular notions of UFOs. Artists, he wrote, "along with everyone else, became familiar with the UFO phenomenon whether they were curious about it or not."

Hopkins postulated that the image of the flying saucer had become an enigmatic icon in humanity's collective cultural consciousness and that it now served as a powerful symbol in artistic expression. He found support for this theory in the writings of the renowned psychologist Carl Jung, who some years earlier had associated the circular craft with a mandala, a universal symbol signifying "the wholeness of the self."

Most of the works on this and the following pages, including Chris van Allsburg's whimsical *Event at the Observatory (above),* appeared in a 1982 exhibit called the UFO Show, which Hopkins organized at New York's Queens Museum. Some pieces were inspired by the artists' own UFO encounters; others were not. All of the artists, however, were apparently moved in some fashion by the familiar yet unsettling concept of otherworldly visitors.

Eerie twilight tones and a camera viewfinder perspective lend journalistic immediacy to Joel Sokolov's Zanesville UFO. The artist says that the 1981 mixed-media composition was inspired by a newspaper photograph of a reported UFO sighting in the Ohio town of Zanesville.

Jene Highstein's 1977 concrete sculpture titled Flying Saucer seems almost to hover amid the tall grass in a field near Park Forest, Illinois. Budd Hopkins, in his notes for the UFO Show, wrote that the "absolute closure and absence of detail" of the massive work "suggest magically charged interiors, imposing and impenetrable."

With its large blank eyes and otherwise featureless head, Terry Rosenberg's untitled 1982 cowhide sculpture evokes in abstract form the concept of the extraterrestrial as an ominous being. The bottom of the figure seems to melt into a void, symbolizing, perhaps, the difficulty some purported UFO abductees have in remembering the lower torsos of their alien captors.

In their 1978 composition UFO Attack: Building in Flames, artists Peter Robbins and Saul Ostrow portray an alien spacecraft committing a malevolent act against the world's workaday people, as represented by the old-style factory. The violent and nightmarish quality of the painting seems to suggest the dangers inherent in the unknown. Robbins's contribution to the work may have been based on the UFO that he claims to have seen at the age of fourteen.

A seven-foot-tall sculpture in mixed media and found objects, Susan Woldman's 1990 Starlady reflects the artist's fascination with otherwordly beings. Woldman reports that she has felt very strongly the force of "alien energies."

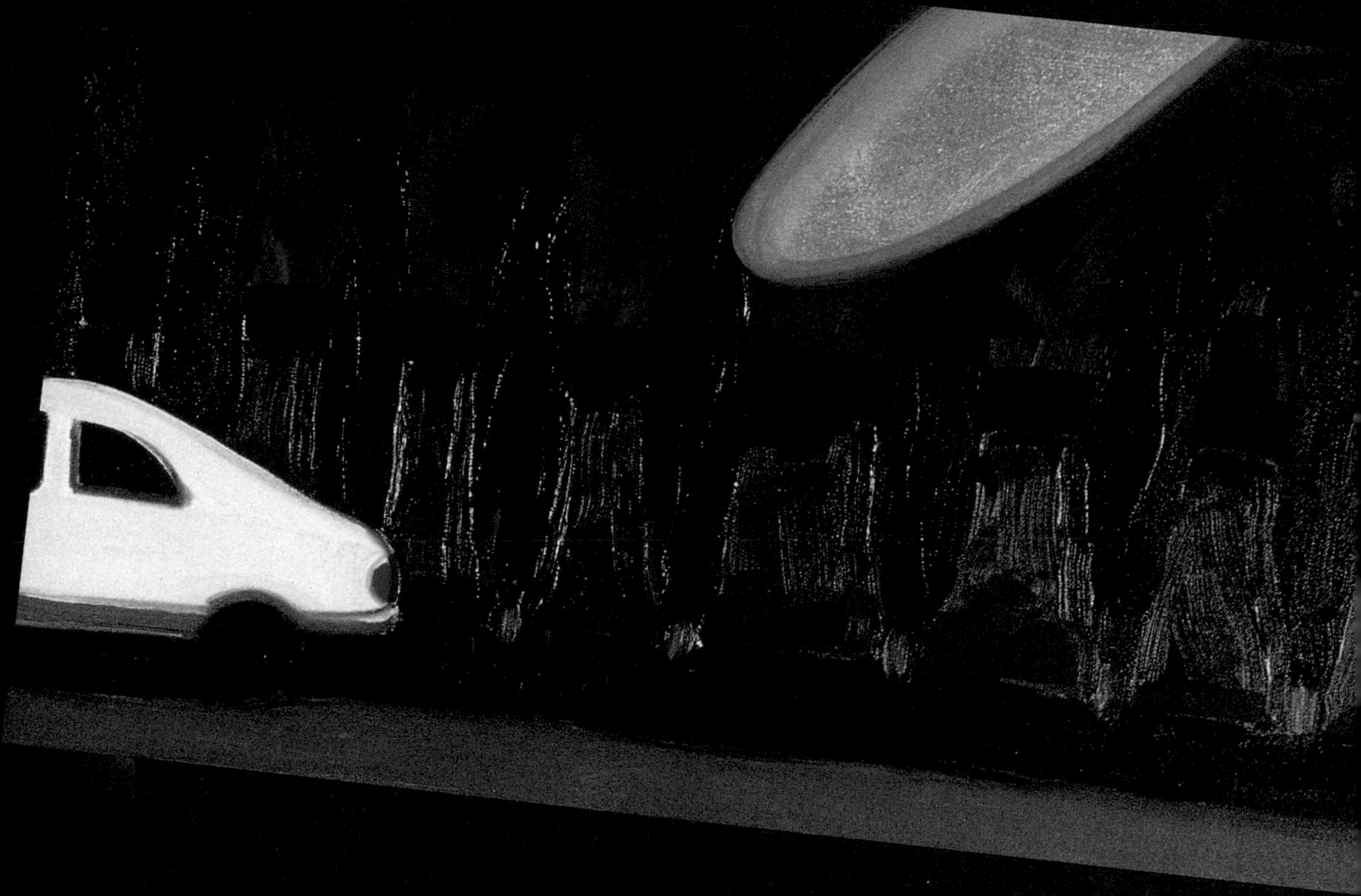

Despite the seemingly puckish double entendre of its title, Round Trip, this 1977 painting by Eve Vaterlaus conveys a frightening moment in a close encounter with an alien spacecraft. The stark tableau invites speculation: The UFO may be pursuing the car, or fleeing after abducting the vehicle's occupants, or even returning home after restoring abductees to the familiar setting from which they were taken.

Intense light emanates from a mysterious underground source while a door opens into a barren desert landscape in Rosemary Osnato's 1982 oil painting entitled Top of the Tunnel. Inspired by an alien encounter that Osnato claims to have experienced in 1973, the surreal scene—in particular the suggestion of an underground world—coincides with details of several other reported abductions.

An alien craft appears to be emitting some unknown kind of energy that is captured by a human figure in this untitled work (above) from 1982 by Keith Haring. The UFOs in Haring's art take a form that is immediately recognizable, even when depicted in the sheer simplicity of his pop-art style.

An ebony disk dominates Budd Hopkins's 1970 work Black and White Lithograph (right). According to Hopkins, the work reflects the influence of his UFO encounter. He wrote that the disk motif came into being, "subtly, more by osmosis than by deliberate decision, yet I never doubted its 'rightness.' "

Two photographs from Dean Cady's Series D collection, made between 1978 and 1980, strike the theme of an extraordinary creature in ordinary circumstances. The bizarre figure, created by a human model wearing a facsimile of an ancient Roman theater mask, hints at a subtle play on words by depicting an alien who has assumed the role of an Earthling alienated from society—lying alone in a cheerless room (right) and eating alone at a diner (above). Reinforcing the figure's otherworldly aspect are the light fixtures in the photograph above, which seem to hover over him like a fleet of UFOs.

ACKNOWLEDGMENTS

The editors would like to thank the following for their assistance in the preparation of this volume:

Walter Andrus, Jr., Mutual UFO Network (MUFON), Seguin, Texas; Don Berliner, Alexandria, Virginia; Thomas E. Bullard, Bloomington, Indiana; Mark C. Curtis, Pensacola Beach, Florida; Raymond E. Fowler, Wenham, Massachusetts; Roger Gaillard, Yverdon, Switzerland; Richard Hall, Brentwood, Maryland; Budd Hopkins, New York; William G. Hyzer, Janesville, Wisconsin; David Jacobs, Temple University, Philadelphia; Dr. Evelyn Klengel, Vorderasiatisches Museum, Staatliche Museen zu Berlin, Berlin; Kevin Kooy, Lethbridge, Alberta, Canada; W. C. Levengood, Grass Lake, Michigan; Betrand Méheust, Mezilles, France; Mark Rodeghier, Center for UFO Studies (CUFOS), Chicago; Carol and Rex Salisberry, Navarre Beach, Florida; Zecharia Sitchin, New York; Robert Swiatek, Arlington, Virginia; Fred Whiting, Mount Rainier, Maryland.

BIBLIOGRAPHY

Academic American Encyclopedia. Princeton, N.J.: Aretê, 1980.
Aksvonov, L., and Boris Zverev, "Aliens Visit Voronezh." *Moscow News,* no. 43, 1989.
Ashtar and Sananda ("received thru Tuieta"), *Conclave: Meeting of the Ones.* Sedona, Ariz.: Portals of Light, 1989.
Asimov, Isaac, *Extraterrestrial Civilizations.* New York: Crown, 1979.
"Blowing the Whistle on the Government's Cover-up." *UFO Universe,* October-November 1991.
Brosses, Marie-Therese de, "F-16 Radar Tracks UFO." Transl. by Robert Durant. *MUFON UFO Journal,* August 1990.
Bulfinch, Thomas, *Mythology.* New York: Dell, 1959.
Bullard, Thomas E., *UFO Abductions: The Measure of a Mystery* (Vol.1). Bloomington, Ind.: The Fund for UFO Research, 1987.
Burgess, Martin, "The Great Cultures of the Ancient World and the Work of Zecharia Sitchin." *FSR Magazine,* March 1991.
Cavendish, Richard, *Visions of Heaven and Hell.* London: Orbis, 1977.
Cavendish, Richard, ed., *Man, Myth & Magic* (Vol. 2). New York: Marshall Cavendish, 1985.
Chorost, Michael, *The Summer 1991 Crop Circles: The Data Emerges.* Washington, D.C.: Fund for UFO Research Inc., 1991.
Clark, Jerome, *UFOs in the 1980s* (Vol. 1 of *The UFO Encyclopedia).* Detroit: Apogee Books, 1990.
Collier's Encyclopedia (Vol. 21). New York: Macmillan Educational Company, 1991.
Curran, Douglas, *In Advance of the Landing: Folk Concepts of Outer Space.* New York: Abbeville Press, 1985.
Druffel, Ann, and D. Scott Rogo, *The Tujunga Canyon Contacts.* New York: New American Library, 1980.
Ecker, Don, "Mankind's Cosmic 'Creators.' " *UFO,* January-February 1992.
The Encyclopedia Americana (Vol. 26). Danbury, Conn.: Grolier, 1991.
Evans, Hilary, "Abductions: A Position Statement." *Journal of UFO Studies* (Chicago), March 1989.
"The Fall of a Warlock." *Time,* April 2, 1979.
Farkas, Ann E., Prudence O. Harper, and Evelyn B. Harrison, eds., *Monsters and Demons in the Ancient and Medieval Worlds.* Mainz am Rhein: Philipp von Zabern, 1987.
Farson, Daniel, *Vampires, Zombies, and Monster Men.* Garden City, N.Y.: Doubleday, 1976.
Faruk, Erol A., "The Delphos Case: Soil Analysis and Appraisal of a CE-2 Report." *Journal of UFO Studies* (Chicago), March 1989.
Fein, Esther B., "Fact, Not Fiction, Tass Says of U.F.O." *New York Times,* October 11, 1989.
Fielding-Hall, H., *Margaret's Book.* London: Hutchinson & Co., 1913.
Fowler, Raymond E.:
The Allagash Affair. Wenham, Mass.: Woodside Planetarium and Observatory, 1990.
The Andreasson Affair. Englewood Cliffs, N.J.: Prentice Hall, 1979.
The Andreasson Affair Phase II. Englewood Cliffs, N.J.: Prentice Hall, 1982.
Fowler, Raymond E., ed., *MUFON Field Investigator's Manual.* Seguin, Tex.: MUFON, no date.
Froud, Brian, and Alan Lee, *Faeries.* New York: Harry N. Abrams, 1978.
Garelik, Glenn, "The Great Hudson Valley UFO Mystery." *Discover,* November 1984.
Gods and Goddesses (The Enchanted World series). Alexandria, Va.: Time-Life Books, 1986.
Good, Timothy, *Above Top Secret.* New York: William Morrow, 1988.
Haines, Richard F., "Analysis of a UFO Photograph." *Journal of Scientific Exploration,* Vol. 1, no. 2, 1987.
Hall, Richard, *Uninvited Guests: A Documented History of UFO Sightings, Alien Encounters & Coverups.* Santa Fe, N.Mex.: Aurora, 1988.
Holt, Alan C., "UFO Light Beams: Space-Time Projections." MUFON Symposium Proceedings, Fifteenth Annual MUFON UFO Symposium, San Antonio, Tex., July 1984.
Hopkins, Budd:
"An Artist's View of the UFO Phenomenon." The UFO Show, New York: The Queens Museum, August 6-October 24, 1982.
Intruders. New York: Random House, 1987.
Huneeus, Antonio, "Breaking Down 'The Wall' of Eastern Europe's UFO Secrets." *UFO Universe,* June-July 1991.
Hynek, J. Allen, Philip J. Imbrogno, and Bob Pratt, *Night Siege: The Hudson Valley UFO Sightings.* New York: Ballantine Books, 1987.
Imbrogno, Philip J., "The Boomerang Mystery." MUFON Symposium Proceedings, Fifteenth Annual MUFON UFO Symposium, San Antonio, Tex., 1984.
Kiessling, Nicolas, *The Incubus in English Literature: Provenance and Progeny.* Washington State University Press, 1977.
Kinder, Gary, *Light Years.* New York: Atlantic Monthly Press, 1987.
Kramer, Samuel Noah, and the Editors of Time-Life Books, *Cradle of Civilization* (Great Ages of Man series). New York: Time, 1967.
Kren, Thomas, and Roger S. Wieck, *The Visions of Tondal.* Malibu, Calif.: The J. Paul Getty Museum, 1990.
Lehner, Ernst and Johanna:
Devils, Demons, Death and Damnation. New York: Dover, 1971.
A Fantastic Bestiary: Beasts and Monsters in Myth and Folklore. New York: Tudor, 1969.
Matthews, Robert, "Stargazers Feel Pulling Power of Planet X." *Sunday Telegraph,* November 3, 1991.
Meessen, Auguste, "The Belgian Sightings." *International UFO Reporter,* May-June 1991.
Miller, Penny, *Myths and Legends of Southern Africa.* Cape Town, South Africa: T. V. Bulpin, 1979.
Montgomery, Ruth, *Aliens among Us.* New York: Fawcett Crest, 1985.
Randle, Kevin D., and Donald R. Schmitt, *UFO Crash at Roswell.* New York: Avon Books, 1991.
"Remarkable Military Encounter in Belgium." *International UFO Reporter,* July-August 1990.
Rodeghier, Mark, "Roswell, 1989." *International UFO Reporter,* September-October 1989.
Sagan, Carl, et al., *Murmurs of Earth.* New York: Random House, 1978.
Sikes, Wirt, *British Goblins: Welsh Folk-lore, Fairy Mythology, Legends and Traditions.* Wakefield, Yorkshire: EP Publishing, 1973.
Sitchin, Zecharia:
Genesis Revisited. New York: Avon Books, 1990.
The 12th Planet. New York: Avon Books, 1976.
The Wars of Gods and Men. New York: Avon Books, 1985.
Spencer, John, and Hilary Evans, eds., *Phenomenon: Forty Years of Flying Saucers.* Arlington, Va.: Central Library, 1988.
Steiger, Brad, and Francie Steiger, *The Star People.*

New York: Berkley Books, 1981.
Strainic, Michael, "Once Upon a Time in the Wheat." *MUFON UFO Journal,* December 1991.
Strieber, Whitley, *Communion.* New York: Beech Tree, 1987.
Stringfield, Leonard H., *UFO Crash/Retrievals: The Inner Sanctum.* Cincinnati: Leonard H. Stringfield, 1991.
Swords, Michael D.:
Introduction to "Issues Forum: UFO Abductions." *Journal of UFO Studies* (Chicago), March 1989.
"UFOs as Time Travelers." *International UFO Reporter,* September-October 1990.
Teltsch, Kathleen, "U.N. Hears Call to Debate U.F.O.'s." *New York Times,* October 8, 1977.
Templeton, David, "The Uninvited." *Pittsburgh Press,* May 19, 1991.
Thompson, Keith, *Angels and Aliens.* Reading, Mass.: Addison-Wesley, 1991.
The UFO Phenomenon (Mysteries of the Unknown series), Alexandria, Va.: Time-Life Books, 1987.
Vague D'Ovni sur la Belgique. Brussels: SOBEPS, 1991.
Vallee, Jacques, *Messengers of Deception: UFO Contacts and Cults.* Berkeley, Calif.: And/Or, 1979.
Walker, Tom, "Could Be It's Full of Martians Hoping to Get In on 1992." *The Wall Street Journal Europe,* October 5-6, 1990.
Walters, Ed, and Frances Walters, *The Gulf Breeze Sightings.* New York: Avon Books, 1990.
Wells, H. G., "The Things That Live on Mars." *Cosmopolitan,* March 1908.
Wilson, Peter Lamborn, *Angels.* New York: Pantheon Books, 1980.

PICTURE CREDITS

The sources for the illustrations in this book are shown below. Credits from left to right are separated by semicolons; credits from top to bottom are separated by dashes.

Cover: Art by Alfred T. Kamajian. 3 and initial alphabet: Art by Stephen Wagner. 7-15: Art by Wendy Popp. 17: Art by Alfred T. Kamajian. 18: Courtesy Raymond E. Fowler, Wenham, Mass.—drawing by Chuck Rak, courtesy Raymond E. Fowler, Wenham, Mass. 19: Drawing by Charlie Foltz, courtesy Raymond E. Fowler, Wenham, Mass.—drawings by Chuck Rak, courtesy Raymond E. Fowler, Wenham, Mass. (2). 20-21: Jack Weiner, courtesy Raymond E. Fowler, Wenham, Mass. 22-23: Fortean Picture Library, Clwyd, Wales; Michael H. Rogers, Show Low, Ariz.; Fred Youngren, courtesy Raymond E. Fowler, Wenham, Mass. 24: Drawing by Betty Andreasson Luca, courtesy Raymond E. Fowler, Wenham, Mass.; Jack Weiner, courtesy Raymond E. Fowler, Wenham, Mass.; Kathie Davis, courtesy Budd Hopkins, New York. 25: From *UFO Abduction at Mirassol,* courtesy Wendle C. Stevens, UFO Photo Archives, Tucson, Ariz.; drawing by Rick Kenyon, from *UFO Phenomena and the Behavioral Scientist,* edited by Richard F. Haines, The Scarecrow Press, Metuchen, N.J., 1979. 27: Raymond E. Fowler, Wenham, Mass. 28: © Betty Andreasson Luca, Rockfall, Conn. 29: Courtesy Edith Fiore, Clinical Psychologist, Saratoga, Calif. 30-31: © Betty Andreasson Luca, Rockfall, Conn. 33: Douglas Curran, North Vancouver, from *In Advance of the Landing: Folk Concepts of Outer Space,* Abbeville Press, New York, 1985. 34: Raymond E. Fowler, Wenham, Mass.—courtesy Budd Hopkins, New York—courtesy Raymond E. Fowler, Wenham, Mass. 35: Rosemary Osnato, New York. 37: Drawing by Betty Andreasson Luca, courtesy Raymond E. Fowler, Wenham, Mass.—Kathie Davis, courtesy Budd Hopkins, New York. 39: John N., courtesy Budd Hopkins, New York (2). 40: From *An Alien Harvest: Further Evidence Linking Animal Mutilations and Human Abductions to Alien Life Forms* by Linda Moulton Howe © 1989, Huntingdon Valley, Pa. (2). 43: Roger-Viollet, Paris. 44: By courtesy of the Board of Trustees of the Victoria and Albert Museum, London. 45: Mary Evans Picture Library, London; Gianni Dagli Orti, Paris. 46: Jean-Loup Charmet, Paris; from *Myths and Legends of Southern Africa* by Penny Miller, T. V. Bulpin Publications, Cape Town, 1979. 47: Bibliothèque Nationale, Paris—General Research Division, The New York Public Library, Astor, Lenox and Tilden Foundations, New York. 48: Mary Evans Picture Library, London—from *Les Visions du Chevalier Tondal,* "Tondal's Guardian Angel Comes to his Aid," by Simon Marmion (attrib.), illuminator, and David Aubert, scribe, collection of the J. Paul Getty Museum, Malibu, Calif. 49: Giraudon, Paris. 50: Alan Lee, Chagford, Devon. 51: Michael Holford, Loughton, Essex—from *Devils, Demons, Death and Damnation* by Ernst and Johanna Lehner, Dover Publications, New York, 1971. 52-53: Scala, Florence; detail of *Immortals* by Fugai Honko/Los Angeles County Museum of Art, purchased with funds provided by the Barbara Becker estate, the Disney Corporation, the Fluor Corporation, Chevron, William Randolf Hearst Collection, the Anita Baldwin estate, and Los Angeles County Funds. 55: Art by Alfred T. Kamajian. 56-57: Courtesy Philip Imbrogno, Greenwich, Conn. 58: Michel Tcherevkoff/© 1984 *DISCOVER.* 59: Wayne Sorce, Brooklyn Heights, N.Y.; Michael Rowe/© 1984 *DISCOVER.* 60-61: Mary Evans Picture Library, London. 62-63: Sketches by Roma Torshin, courtesy *Moscow News Weekly,* No. 43, 1989. 64: Fred R. Youngren, from *The Andreasson Affair* by Raymond E. Fowler, Prentice-Hall, Englewood Cliffs, N.J., 1979. 66-67: H. McRoberts, Vancouver Island. 68-69: Edward Walters, Gulf Breeze, Fla. 70-71: Mutual UFO Network (MUFON), Seguin, Tex. 74: Edward Walters, Gulf Breeze, Fla. 75: © 1990 Mark C. Curtis/Eric Buckelew, Pensacola Beach, Fla. 77: Kevin Kooy/*Lethbridge Herald,* Alberta. 78-79: Dr. W. C. Levengood, Grass Lake, Mich. 80: © G. Mossay. 81: Belgian Air Force, courtesy of SOBEPS, Brussels. 82-83: Courtesy Inter-Governmental Philatelic Corporation, New York. 84: Drawing by Glenn Dennis, from *UFO Crash at Roswell* by Kevin D. Randle and Donald R. Schmitt, Avon Books, N.Y., 1991. 85: J. Allen Hynek Center for UFO Studies, Chicago. 86: Charles M. Hanna, courtesy Stan Gordon, Greensburg, Pa. 87: Reprinted with the permission of the *Boston Herald*/reprinted with the permission of United Press International, Inc.—reprinted with permission from the *Philadelphia Inquirer*—reprinted with the permission of the *Boston Herald*/reprinted with the permission of United Press International, Inc.; reprinted with permission from the *Philadelphia Inquirer.* 89: NASA. 90-91: Art by Alfred T. Kamajian, based on drawing by Zecharia Sitchin—Oriental Division, The New York Public Library, Astor, Lenox and Tilden Foundations, New York; Museum of Baghdad, from *L'Art de Sumer* by André Parrot, UNESCO/Albin Michael 1969 (Plate #1)/DR; Staatliche Museen zu Berlin, Vorderasiatisches Museum. 92: Art by Alfred T. Kamajian, based on drawing by Zecharia Sitchin—Staatliche Museen zu Berlin, Vorderasiatisches Museum, courtesy Zecharia Sitchin, New York. 93: Art by Alfred T. Kamajian, based on drawing by Zecharia Sitchin—art by Alfred T. Kamajian, based on drawings by Zecharia Sitchin; art by Alfred T. Kamajian. 94: Hermitage Museum, St. Petersburg, Russia—The University Museum, University of Pennsylvania (neg. #S4-140541). 95: Zecharia Sitchin, New York; from *The Tower of Babel* by André Parrot, Philosophical Library, New York, 1955, courtesy Zecharia Sitchin, New York—courtesy of the Trustees of the British Museum, London. 96-97: Réunion des Musées Nationaux, Paris; Zecharia Sitchin, New York; Réunion des Musées Nationaux, Paris. 99: Art by Alfred T. Kamajian. 100-101: Eduard Meier, courtesy Cece Stevens-Nelson, UFO Photo Archives, Tucson, Ariz.; courtesy Cece Stevens-Nelson, UFO Photo Archives, Tucson, Ariz. 102-103: From *A Fantastic Bestiary* by Ernst and Johanna Lehner, Tudor Publishing Company, New York, 1969; illustration by William R. Leigh, photographed from *COSMOPOLITAN MAGAZINE,* Vol. XLIV, March 1908, No. 4, p. 334. 107: Drawing by Chris Blomme, courtesy Michael A. Persinger,

Laurentian University, Sudbury, Ontario; Brian T. Brady, Denver, Colo./U.S. Department of Interior, Bureau of Mines. 108: Cornell University Photo, Ithaca, N.Y. 109: NASA—NASA/JPL (2). 112-113: Peter Kurz/Gamma-Liaison, New York. 116-117: Douglas Curran, North Vancouver, from *In Advance of the Landing: Folk Concepts of Outer Space,* Abbeville Press, N.Y., 1985. 118-119: Courtesy Zecharia Sitchin, New York. 120: © *Flying Saucer Review,* High Wycombe, Buckinghamshire. 122-123: The Bettmann Archive, New York. 124-125: Gary Fong/© *SAN FRANCISCO CHRONICLE,* reprinted by permission. 127: *Event at the Observatory* by Chris Van Allsburg, photo courtesy Budd Hopkins, New York. 128-129: Sculpture by Jene Highstein/photograph by Douglas Curran from *In Advance of the Landing: Folk Concepts of Outer Space,* Abbeville Press, N.Y., 1985; Joel Sokolov, New York. 130-131: *Untitled* by Terry Rosenberg, courtesy Jon Oulman Gallery, Minneapolis; art by Peter Robbins/Saul Ostrow, New York, 1977; sculpture by Susan Woldman, New York. 132-133: Art by Rosemary Osnato, courtesy Budd Hopkins, New York; art by Eve Vaterlaus, courtesy Budd Hopkins, New York. 134: © 1992 The Estate of Keith Haring, New York. 135: Budd Hopkins, New York. 136-137: © 1992 Dean Cady, New York City.

INDEX

Numerals in italics indicate an illustration of the subject mentioned.

E

F

G

H

I

J

K

L

M

N

O

P

Q

R

S

Time-Life Books is a division of Time Life Inc.,
a wholly owned subsidiary of
THE TIME INC. BOOK COMPANY

First printing. Printed in U.S.A.
Published simultaneously in Canada.
School and library distribution by Silver Burdett Company, Morristown, New Jersey 07960.

Library of Congress Cataloging in Publication Data
Alien encounters / by the editors of Time-Life Books.
p. cm.—(Mysteries of the unknown)
Includes bibliographical references and index.
ISBN 0-8094-6545-0 (trade)
ISBN 0-8094-6546-9 (library)
1. Unidentified flying objects—Sightings and encounters.
2. Life on other planets.
I. Time-Life Books. II. Series.
TL789.3.A44 1992
001.9'42—dc20 92-13265
CIP

Other Publications:

LOST CIVILIZATIONS
ECHOES OF GLORY
THE NEW FACE OF WAR
HOW THINGS WORK
WINGS OF WAR
CREATIVE EVERYDAY COOKING
COLLECTOR'S LIBRARY OF THE UNKNOWN
CLASSICS OF WORLD WAR II
TIME-LIFE LIBRARY OF CURIOUS AND UNUSUAL FACTS
AMERICAN COUNTRY
VOYAGE THROUGH THE UNIVERSE
THE THIRD REICH
THE TIME-LIFE GARDENER'S GUIDE
TIME FRAME
FIX IT YOURSELF
FITNESS, HEALTH & NUTRITION
SUCCESSFUL PARENTING
HEALTHY HOME COOKING
UNDERSTANDING COMPUTERS
LIBRARY OF NATIONS
THE ENCHANTED WORLD
THE KODAK LIBRARY OF CREATIVE PHOTOGRAPHY
GREAT MEALS IN MINUTES
THE CIVIL WAR
PLANET EARTH
COLLECTOR'S LIBRARY OF THE CIVIL WAR
THE EPIC OF FLIGHT
THE GOOD COOK
WORLD WAR II
HOME REPAIR AND IMPROVEMENT
THE OLD WEST

For information on and a full description of any of the Time-Life Books series listed above, please call 1-800-621-7026 or write:
Reader Information
Time-Life Customer Service
P.O. Box C-32068
Richmond, Virginia 23261-2068

This volume is one of a series that examines the history and nature of seemingly paranormal phenomena. Other books in the series include:

Mystic Places
Psychic Powers
The UFO Phenomenon
Psychic Voyages
Phantom Encounters
Visions and Prophecies
Mysterious Creatures
Mind over Matter
Cosmic Connections
Spirit Summonings
Ancient Wisdom and Secret Sects
Hauntings
Powers of Healing
Search for the Soul
Transformations
Dreams and Dreaming
Witches and Witchcraft
Time and Space
Magical Arts
Utopian Visions
Secrets of the Alchemists
Eastern Mysteries
Earth Energies
Cosmic Duality
Mysterious Lands and Peoples
The Mind and Beyond
Mystic Quests
Search for Immortality
The Mystical Year
The Psychics